ECOSYSTEM RESEARCH IN SOUTH AMERICA

BIOGEOGRAPHICA

Editor-in-Chief

J. SCHMITHÜSEN

VOLUME VIII

DR. W. JUNK B.V., PUBLISHERS, THE HAGUE 1977

ECOSYSTEM RESEARCH IN SOUTH AMERICA

Edited by Paul Müller

with contributions by

H. Lamprecht, U. Irmler, V. Quintanilla, H. Jakobi,
O. Fränzle, J. Lescure and F. Christiansen-Weniger

DR. W. JUNK B.V., PUBLISHERS, THE HAGUE 1977

ISBN 90 6193 209 2

CONTENTS

STRUCTURE AND FUNCTION OF SOUTH AMERICAN FORESTS

H. LAMPRECHT

Abstract

The study deals mainly with the tropical evergreen forests and moist forests of northern South America. The basal area of these intensely mixed natural forests with many species reaches its minimum in the tropical tropophytic forest and its maximum in the Andean cloud forests. The curve of diameter distribution is typical of a selection forest. As far as protective functions are concerned these forests are satisfactory, but their productive functions are insufficient. The task of forestry is to convert these unproductive natural stands into productive commercial forests without decreasing their protective and social functions.

With a forested area of a billion ha., an average of 45 % of the surface area forested and 5 ha. of forest for each person in the population, Latin America is the most densely forested region of the world. The Latin American countries have about $\frac{1}{4}$ of the total forested area of the world KNIGGE (1972). The forests of South America still cover today about 880 million ha., divided by HUECK (1966) into 40 different forest regions. A further 242 million ha. appear in the stitistics as bush land. In view of this enormous extent and complexity it is not possible in an hour to give a complete outline of the structure and function of South American forests.

I will limit myself, therefore, to a few examples from the tropical zone and mainly from northern South America. The southern forested region will not be considered, but since Mr. QUINTANILLA is discussing the Chilean forests in his report, in this context the most important area of forest in southern South America, this gap will be largely closed. Because of time I must further limit discussion to evergreen rain forest and to rain-green moist forests. As closed forests these occupy 540 million hectares in tropical Latin America, and 380 million hectares as open tropical woodland and bush formations. With a total of 800 million hectares South America has the greatest share of the forests of this tropical zone. In tropical Africa there are 660 million ha. and in Asia 370 million ha. It is, therefore, apt to describe South America as the continent of forests. These facts above are sufficient grounds to concern oneself with the structure, composition and function of the forests. Other points can be made in the following text.

The forests of Neotropis impress the observer first in their diversity of species, especially of trees, and then in their complicated stand structure. CAIN *et al.* (1956) counted 216 plant species in 2 ha. of an undisturbed forest near Belém. The life form spectrum consisted of 157 phanerophytes, 30 lianes, 19 epiphytes, 2 chamaephytes, 6 hemicryptophytes, 2 cryptophytes, and 0 therophytes. KLINGE & RODRIGUES (1968) counted in a terra firme forest near Manaos, on a 2000 m^2 quadrat, 505 plant species.

The following consideration will be concerned with woody plants having a BHD* of 10 cm and more, unless otherwise remarked. The results are based upon
* BHD = breast height diameter (1.3m)

1

probes with a surface area of 1 ha. Since the minimum area, based fairly reliably on a species-area survey, in general is between 2,500 – 10,000 m², the probes can be taken as representative. As basic data the abundance, frequency, dominance and vertical distribution of individual tree species were measured. With the help of these values the floristic structure of a stand can be quantitatively clearly defined LAMPRECHT (1969b). In the appendix the results of structural analyses of evergreen rain forest, Andean evergreen cloud forest and the rain-green moist forest are given together for elucidation and to assist in the following presentation. Before discussing particularities I wish to give a general outline of the tree species to be expected in the various forest formations according to the climatic conditions, selecting as example Venezuela. According to the forest atlas MAC (1961) it is possible to reckon the following minimum number of tree species for the whole country.

Minimum tree species in different Venezuelan forest types

rainfall	average annual temperature 0°C (relative to altitude above sea level in m.)		
mm per annum	> 24 (0-400)	12-24 (400-2300)	6-12 (2300-3300)
2000 - 4000	172 tree species 9 palm species	116 tree species 5 palm species	99 tree species
1000-2000	178 tree species 8 palm species	190 tree species 6 palm species	2 palm species
500-1000	36 tree species no palms		

Obviously, at best, these values only provide a crude outline, but they showed clearly that the number of species diminished on the one hand with diminishing rainfall associated with a longer and more marked dry season, on the other hand with altitude associate with falling average annual temperature. The sole exception is the semi-evergreen moist montane-forest with the highest number of tree species, but this can be easily explained in that here elements from the lowlands and from the mountains, form the evergreen forest and from the dry cool forest, mingle.

Initiatory research by VEILLON (personal communication) also concerned with the relationship between species numbers and rainfall in the Venezuelan tierra caliente forest type gave the following values: 100% species in evergreen rain forest, 67% in rain-green moist forest, 40% in rain-green dry forest, and 12% in thorn bush. A number of ascents into the cloud forest girdle of the Venezuelan Andes revealed the following results LAMPRECHT (1958):

    100%   (56 species)   heights u.s.l. 2200 - 2600 m.
    68%   heights u.s.l. 2600 - 2800 m.
    30%   heights u.s.l. 2800-3200 m.

FORSTER (1973) discovered in the rain forest of the central Magdalena valley on 3 ha. in total 103 tree species, and on each hectare between 70-85 species with 571, 645, 657 trees per hectare. RODRIGUES (KLINGE & RODRIGUES 1968) came to similar figures in the Manaos region. A large inventory for 137,000 ha. gave 470 species of over 25 cm BHD with an average of 65 (50-80) species per ha.

2

The number of trees was on average 102 per ha.

In the rain-green moist forest of the Venezuelan llanos BERNAL (1967) found 82 species over 20 cm BHD on 100 ha. According to FINOL (1969) and own surveys, the number of species over 10 cm BHD fluctuates between 40 and 50 per hectare, and is therefore clearly less than the comparative rain forest. The same applies to the tree number, in rain-green moist forest not exceeding 300-400 trees per hectare.

A rough transect to establish the mixing intensity produced the so-called mixture quotient, that is the number of species to the number of trees. In the rain forest as in the rain-green moist forest this value is of the order between 1 to 5 and 1 to 10, that means each species on average is present only as 5-10 individuals per hectare. Such extreme proportions as suggested at first sight by the mixture quotient are rarely reached. More often in every probe 40-50% of the trees belong to 10-15% of the species. A comparatively small group dominates the local species spectrum, and the bulk of the other species together comprise the remainder, often very much in the background.

Naturally the importance of a species in the arrangement of a stand is not solely determined by the number of individual of that species, but also upon space they occupy in the community. This measure is the so-called dominance or degree of cover, that is the horizontal projection of the total leaf and shoot system of a species on the soil suface. For trees it is more correctly the sum of the projections of individual crowns. In tropical forest measurements of crowns of this type are difficult and time-consuming so that dominance is generally estimated in terms of the sum of basal areas of each species. According to specific probes made by LAMPRECHT (1972) the values in northern South America are:

     lowland raing-green moist forest   15-30 m²/ha.
     lowland rain forest   30-40 m²/ha.
     mountain rain forest (cloud forest) 40-60 m²/ha.

This basal area, apart from possible clearance, is clearly different for various forest formations. The hypothesis developed by BRÜNIG (1968) that in the primaeval forest the basal area is an indication of the natural fertility of the site cannot be here discussed. Nevertheless, it is interesting to note that not lowland rain forest, as might be expected, but cloud forest of high altitudes from 2,000 to 2,400 metres shows the highest basal area value. Without attempting to enter into explanations it may be mentioned that in the ecological classification after the biotaxonomic index of VARESCHI (1968) mountain cloud and rain forest are plainly at the top as optimal forests. If the dominance according to species is analysed so it is found again that 10-15% of all the tree species comprises 50-70% of the total basal area. This is not a question of a high tree frequency but often associated with a few particularly massive individuals.

A further essential structural element is the vertical and horizontal distribution of individual species in the stand. The horizontal distribution is found by the calculation of total frequency, the vertical distribution by frequency in the different crown levels. Throughout in the frequencies of the whole stand the lowest class of values predominate. Generally well over half the species have values under 20%. To the well known fact that the South American forests are extraordinarily mixed must be added the knowledge that the composition of the mixtures varies markedly locally. Of course in both the rain-green moist forest and the evergreen forest there are the so-called horizontally well distributed species, that is species with high mixing values in that they are found in all mixed stands. Generally but not invariably they are species also found in high abundance. With additionally greater dominance they play a dominant role in the arrangement of a stand. The horizontally well distributed species usually represent 5-15% of the total number of species.

Vertically the forest types under consideration can be divided in a rather schematic way into three canopy or crown horizons or layers of trees over 10 cm BHD. In the upper horizon there are 10-20% of the total number of trees, in the middle horizon 25-50%, and in the lowest horizon 40-60% (on average 50%).

According to tree species the following points can be made.

1. Not all species are present in all horizons. Generally the following can be distinguished:

a. proximate species, that is trees of the 2nd and 3rd scale order, exclusively limited in occurrence to the middle and lowest layers.

b. more or less exclusively uppermost-layer trees, in practice only present in the highest canopy horizon.

c. vertically well-distributed species that are more or less regularly found in all canopy layers. Generally those form about 15-20% of the total number of species. Normally the most abundant and often the horizontally well-distributed species belong to this group.

2. The most botanically-complete spectrum of species is shown by the middle and lower layers. The upper layer in this respect is the poorest, but because of the small number of individuals present it shows in a 1:3 mixing quotient the highest degree of mixing. It follows also that the middle and lowest layers, considered separately, are more strongly mixed than the whole stand.

A further essential characteristic of the structure is unquestionably the distribution of diameter sizes. In undisturbed primaeval forest of areas often under 1 ha. there is normally a curve similar to a selection forests worked by the removal of occasional individual trees. This is due to the presence of many small trees, and with increasing diameter a falling number of individuals concentrated in the upper layers. This type of curve shows that natural regeneration secures the permanence of the stand. Often it is found that such vegetation seen as a whole is in stable equilibrium, but in no way static.

It is much less easy to interpret analyses of the diameters of individual species, and this is particularly true when the majority of the individuals belong not to the lowest but to higher and more massive classes and therefore appearing exclusively or almost exclusively in the uppermost layer. This phenomenon leads immediately to the question how species with such an irregular distribution of diameter sizes can persist in the stand, when there are obviously not reserves, or only slender reserves of young trees in a waiting stage, prepared for the occurrence of a gap in the upermost layer. This is a fascinating question, and extraordinarily important for any future rational economic usage of the South American forests. The attempt to answer this question, therefore, far exceeds the limits of a static structural analysis, leading automatically to problems to the dynamics of natural tropical forests.

Nevertheless, there is a question that I do wish to examine shortly concerning directly the life sequence of a stand and therefore related to the above question. It is the 'mechanism' by which natural regeneration is achieved in the primaeval forest, and about which, to my understanding, reliable information is still lacking.

It is known that most tree species generally set abundant fruit, and that fresh seeds have a high germination value, often completely lost after a short period of time (e.g. *Tabebuia* sp.). In a number of species (e.g. legumes) the seed coat is hard, making germination difficult or even almost impossible. On the other hand they afford a protection against many dangers. Certainly large quantities of seeds are eaten or destroyed by various animals. Nevertheless in general regeneration does continue to a larger or smaller degree, although less in gaps in a stand which within a short time are covered by bamboos, herbs and lianes. Possibly most species cannot tolerate shade, and die off in a short time, but are continually replaced by new seedlings. Support for this hypothesis as seen in the fact that in the layer 10-100 cm regeneration is frequent, in the layer 101-300 cm much less, but above 300 cm to a BHD of 9.9 cm a certain increase in frequency can be observed. This is possibly a form of enrichment at this layer due to the arrival of survivors from each regeneration wave. From this layer are recruited the quite isolated waiting plants that sooner of later get a chance to grow upwards. Often nature is profligate in seeds and seedlings in order that one or other individual reaches the upper tree-layer. The natural mechanism is succesful in that by this gradual replacement of tree by tree over a long time period a renewal of the stand is achieved and an unlimited survival value for the whole assemblage. The forest is thus, on smaller surface areas, eternal. Beginning and end applies only to the individual trees. It is because of these facts that the forests of South American tropical mountains present such an outstanding protection against erosion. Developments are quite other if catastrophes of some form seriously disturb the normal life sequences. Regeneration analyses according to species show that generally it is the worthless accompanying species rather than the valuable timber trees such as *Cedrela* and *Swietenia* etc. that regenerate. The regeneration process and life sequences of the valuable trees still present problems to the forester.

The key words erosion protection and timber trees introduce the second theme to which belongs wood production and protection effects in the widest sense, the most important tasks not only for today and also in the future for South American forest. A functional analysis of the South American tropical forests begins with the division accepted in old industrial lands: production, protection, recreation. The scale of values is the needs and well-being of man. In the large unoccupied regions as can still be found in South America the forests are considered without function and therefore worthless. Apart from the fact that such regions are often the final refuge of the last surviving autochthonous peoples it may well be that the population explosion and the advance of civilisation will lead in the near future to the opening-up of such untouched territories. It must be considered, however, that the ecological and economic effect of a forest often reach much further afield than is initially apparent. The presence or absence of a forest cover has a marked effect on the water budget of a catchment area. In fact, this is detectable in the complete catchment area from the springs to the river mouth. Forests completely without significance, that is without function in the strictest sense of the word, scarcely exist any more in South America, and will be even less in the future. Naturally there is a gradual scale with reference to their actual importance for the needs and the well-being of the population. Nevertheless in general it can be said completely functional forests become a necessity for South America in view of the population explosion and of increasing industrialisation.

What are the abilities of the South American tropical forests to fulfil production, protection and recreational functions?

Wood and other forest products are usable if the cost of obtaining same is less than, or at least not more than, the sale price. The distances involved, up to which exploitation still pays, depends upon accessibility, that is the transport means available, and also upon the market price of the wood. Generally firm guide lines are not possible, but it is sure that wide areas of the Amazonian forest, the forest of the Orinoco basin, Guayana and Surinam are still beyond the boundary of economic usage. With increasing general accessibility the economic null boundary, where expenditure and takings balance, for more and more species of trees, retreats further into the forest.

So far a few quite incomplete general economic points have been considered. These points will not be further investigated in that, in the present context, interest centres on how far the natural structural forms influence the fulfilling of the production function. The answer is as clear as disappointing. The build and composition of South American tropical rain and rain-green moist forests give a production which is the opposite to that desired by modern wood-using industries. The wood industries requiere large quantities of unvarying homogeneous raw material whilst the tropical forests produce, truly in very large quantities, from the mechanical and technical view point, a great variety of trees with a range of dimensions from twigs to giants. Many cannot be marketed as they are available in too small a quantity or too variable amounts, or because their characteristics are unknown or too poor in view of the quality of the trunk. More expressive than many words is the fact that the extraction of commercially-valuable woods from the Amazon forests is estimated as 1 m^3/ha. According to the latest FAO statistics (FAO 1971) the South American forests in 1969 produced 213 million m^3 of wood without bark, of which 184 million m^3 was firewood. In relation to the total forested area or nearly 800 million ha. this gives a product of $\frac{1}{4}$ m^3 per ha. p.a. It is to be admitted that the statistics are far from complete since there is much unrecorded extraction for the needs of local peasant farmers. Nevertheless the fact cannot be disputed that the direct economic usage of the South American tropical forests is surprisingly light. This is scarcely suprising since the authorities and the population regard the forest as unused or as a serious barrier to an advance in civilisation or at best as potential reserves for the extension of cattle rearing and arable farming. This leads even the forestry people to regard use of accessible forests as a robber economy of the most valuable woods, almost as a form of mining. That the basis of future production is damaged or destroyed is scarcely considered since the economic production capacity up to the present is seen as hopelessly unused. The market value of wood destroyed by the traditional slash and burn primitive semi-nomadic farming has not been assessed. According to an estimate by the FAO in 1960, during a single year some 10 million hectares of forest are felled for this purpose. Since the clear felled and burnt stands are considered of no economic interest their destruction is more or less passively accepted.

Naturally many individuals are pondering how in future the increasing home and international demand for wood can be met by at the same time a rapidly diminishing forest area. Many, including forestry people, have come to the conclusion that the solution of this problem is the establishment of highly productive plantations consisting of one or a few rapid-growth tree species. It is hoped that by these methods on comparatively small areas the wood demand of

6

the market can be met. If this is so the role of the natural forests as wood producers can be relinquished.

Typical of the sketched mode of thought are many of the developments observed today in Brazil. In the south of the country mainly private forest plantation, with state assistance, literally shoot out of the ground, and consist of extensive equal-aged stands of tropical and sub-tropical conifers, eucalypts, and occasionally *Cunninghamia lanceolata*. In the north in contrast the destruction of the Amazonian forests, now also with a massive state participation, increases in tempo. A strong impulse for this clearance derives from the now-inbuilding 'Transamazonica'. Once finished the planned colonisation of small farmers, large farmers, and large ranching concerns, and also definitely the uncontrolled settlements, will in a short period lead to the disappearance of all the easily reached primaeval forest. How rapidly the unleashed colonisation takes place is shown by the so-called **pioneer-front** in Maranhâo state. WEIDELT (1968) found that in the region of the Rio Pindaré six years after the opening of the BR 22, westward towards the primaeval forest the stands up to 15 km distance had been destroyed. Individual outposts of settlement were already 36 km from the road. Similar observations were made by PETRICEKS (1959) immediately after the building of the 'Panamericana' that led in Venezuela from the foot of the Andes to Lake Maracaibo through untouched primaeval forest. In the event the primaeval forest was destroyed.

Quick and more complete results than the clearing activities of individual settlers are to be expected with the massive introduction of modern techniques as now in the Amazonian region. This may particularly be so if, as according to a press report of a few days ago FAZ (14.4.73), in addition to the caterpillars and bulldozers, the defoliants and arboricides employed in Vietnam will also be used to destroy the Brazilian jungles.

Yesterday's dream to create from the 'green hell' of the Amazonian forests an Eden of unparallelled fruitfulness, appears once more a reality if not for today, nevertheless for tomorrow. In professional circles the view persists that future hopes lie in techniques not dedicated to a war to the death with nature. More important in the short term are the effects of the gigantic deforestation taking place not only in Brazil but in South America generally and in the whole tropical world. The first manifestations of these effects are already visible.

With this the second group of functions is touched upon, the protective functions of the forest in the wides sense. It must first be asked if the natural tropical forest by its mere existence through automatic environmental effects fulfills a protective function. This question must clearly be answered with a yes. Tropical rainforest and rain-green forests, because of their already mentioned persistance, and because of the structural characteristics that render them unsatisfactory as timber produce, are ideal protective forests. I can in illustration of this there briefly select only two points. They concern the complex forest as soil protector and as water conservator.

The most important economic protective function of tropical forest may be the conservation of the soil and of the continued productivity of the ecosystem. Investigations by SUAREZ DE CASTRO & RODRIGUEZ-GRANDAS (1961) at Chinchina, Columbia gave the following values of nutrient loss (N, P, K, Ca, Mg) per hectare per annum.

bare soil	440 kg
pasture	64 kg
old coffee plantation with a forest-like structure	7 kg

Still larger masses are lost from forest-free areas by unrestricted soil wash. In Chinchina on pastures this was 1075 kg/ha. per annum, on fallow 1867. In addition to chemical impoverishment a deterioration in the physical proportions of the soil is a direct consequence of exposure, particularly in view of the extremes of the tropical climate. The most striking consequences of the removal of a many layered cover may be soil erosion. According to the research of Suarez de Castro and Rodriquez-Grandas already cited the loss of soil per annum per hectare is

forest-like coffee plantation	0.25 ton
hay meadow	26.64 ton
fallow after 2 maize crops	860.72 ton

It is estimated that in Columbia alone, each year, some 200,000 hectares are lost by soil erosion. For over four fifths of this loss the primitive slash and burn farming is responsible. According to REHM (1973) poor cultivation had already reduced 1.7 billion hectares of the original tropical evergreen and seasonal forests to a useless hard-grass savanna.

Without doubt the reduction of primary forest with its protective capacity to a short-period agriculture and thence to practically productionless waste, as for example the eroded robber landscapes, is one of the greatest problems of tropical land use. It has not been solved, certainly not in South America.

In no way less important in the consequences is the influence of tropical forests on the water budget. Most significant in consideration of balance are the values interception, evapotranspiration, run-off above and through the soil, water infiltration, and water retention capacity of the soil. Also in references to the problem-area forest-water I must limit myself to a few points. The small amount of research available permits the estimate that lowland rain forest intercepts about 15-25% of precipitation. In mountain rain forest the interception may be higher, in the remaining forest formation lower, and apparently lowest in rain-green dry forest. In tropical forests, particularly in rain forests, evaporation is often less than 1% of precipitation but transpiration is higher than in all other vegetation forms. The annual evapotranspiration in rain forest of the 'Tierra Caliente' is about 1300-1400 mm. Similar although mostly lower values are achieved by other vegetation covers. The vegetation of the humid tropics, because of the continuous water requirement of very tall plants, with great structural and floristic differences, has a large water consumption. It is interesting to note the findings of AUBREVILLE (1971) that evapotranspiration in equatorial rain forest approaches the evaporation from an open water surface. In this respect the rainforests of the tropics are comparable with a large sea, and their destruction may have consequences similar to those that would follow from the drying up of a water surface of like dimensions. In areas of rain-green forest with less precipitation it follows that the evapotranpiration is less. There is nevertheless the point that with the decrease in the absolute precipitation the evapotranspiration of forests reaches a relatively higher value compared with other vegetation covers.

The water balance is further influenced by the special faculty of the forest, as the tallest and most richly structured form of vegetation, to abstract moisture from the air. One of the best known examples is the forest of Fray Jorge in north Chile, in which according to KUMMEROW (1966) the mist precipitation is ten times as great as the annual rainfall in extreme cases. Considerable moisture is gathered from mists in the cloud forests, the climax vegetation at high altitudes. Water vapour in the hot moist air rising from the lowland condenses, and the forests are usually enveloped in mist for at least an hour every day. Apart from such and similar cases it would not be anticipated that forests provide more water than unforested catchment areas. In fact in the tropics the converse is the case, as shown by initiatory research of GOUJON (1968) in Madagascar. He found that the water provided bv a

natural forest	=	4.5% of the precipitation
eucalyptus stand	=	4.2% of the precipitation
dry rice field	=	31.4% of the precipitation.

In the rice field, however, the annual soil loss was 6 ton/ha., but erosion was not visible in either the natural forest or the plantation. Such a high water supply may be too expensively bought if linked with soil erosion. In addition it must no be forgotten that it is not the total quantity of water that is of economic significance, but next to quality the regularity and constancy of the supply. Of particular relevance in this context are the streams and stream systems which, because of climatic conditions, show extreme fluctuations in level, as is the case to an outstanding degree with many American rivers. Numerous investigations all over the world have repeatedly confirmed that a forest cover is more effective than any other vegetation in smoothing the irregularities of a water budget. High water is comparatively reduced and drawn out, and in the dry period the level falls, later and more gradually than occurs with streams in forest-free catchments. There is no shortage of South American examples. Unfortunately, however, there are no indisputable quantified examples of changes in water regime after deforestation, but the fact that change do occur in the South American region can scarcely be questioned.

Obviously with a rise in level of high water the erosion danger also increases. With a more marked and longer low water the problems deriving from water shortage are magnified, and at a season of the year when climate limits water supply.

The potential storage facilities of forest soils under tropical conditions was demonstrated for example by KELLMANN (cited by Brünig 1971) in Mindanao. He found that the annual surface flow amounted in a

primary forest	13 mm
savanna	130 mm
rice field (after 12 years use)	498 mm

With these values for the savanna and the rice field the peak surface flow could not be measured owing to the large volumes of water over a short period. A South American example that can be cited is due to SUAREZ DE CASTRO& RODRIQUEZ-GRANDAS. In the old coffee plantation of Chinchina 2.2% of the precipitation is surface flow, in a pasture 30%, and in an harvested maize field 59%.

The faculty of forest soils to absorb and store precipitation, and to release it gradually, due to a quite particular structure, is most important and essential in tropical conditions where other natural regulators are absent. This contrasts with the conditions in high latitudes where the winter snow fall has an important storage function.

These examples may serve to confirm that the forests of the South American tropics by their mere existence fulfill an essential protective role. They may also have shown further that effects of the forests on the tropical environment have from many viewpoints life significance for the present and even more for the future of the inhabitants. With the growth of population and industrialisation the importance of the protective function of wooded areas increases. The forests in South America, without question, are of increasing importance as additional guarantees of living conditions worthy of man.

In contrast to the high protective value the natural forests of tropical America are not suitable for recreation. Apart from their relative impenetrable character, all who know the structure and composition of such forests will readily understand their unattractiveness for recreation and walking. A tourism of primaeval forests, as offered by a few individual travel agencies, are at least provisional for only very limited circles in North America and Europe. The average South American sees nature only from the great city and is not so curious. He fears the 'green hell', not always without reason.

In sum, therefore, the natural forests of South America in the tropical zone because of their structure, composition and life sequence fulfill satisfactorily or outstandingly a protective function, but do not reach high levels in the production of raw materials either quantitatively or qualitatively, and completely fail to fulfill a recreational role. This inability of the tropical forests of South America to meet a balance of needs, and which is determined primarily by structure, presents the tropical foresters with an extraordinarily complex problem. The recent emphasis on plantations to produce timber, although strongly supported with propaganda in forestry circles, offers no true escape from the dilemma. Plantations for wood provision present, without question, an interesting system for the production of large quantities of uniform wood in a short rotation period, but the production is with advantage limited to mass goods. For high value woods the plantation is unsatisfactory. Even more important, from a variety of reasons stemming from systems aspects LAMPRECHT (1969) artificial monoculture is not, or only partly able to meet the essential need for soil protection. For the rest, plantations are made where for this or that reason a forest cover appears necessary, but the exceptions applying to such undertakings are so large that a realistic discussion is not possible. In any case it would be a serious error if the South American foresters wished to limit their activities to the siting and administration of timber plantations.

On the other hand it would be illusory to believe that protective forests, so essential and so important to life, can be preserved simply because of their protective function. The increasing population will inevitably lead to the occupation of such forests by land-hungry peasants or large-scale enterprises and to the destruction of the forests, if it is not demonstrated that they have an additional function as provider of raw materials, and therefore work for the local population. In the South American tropics the optimal future forest is none other than that, which because of its composition, build and development processes, is suitable for a permanent multi-usage. In other words when the retention of the forest is recognised as necessary a change must be made to a natural stocking of higher-value economic stands capable of a yield without a loss of protective value. The decisively important question of the possibility of such a reconstruction must remain open here, since the attempt to find an answer must be made in each

10

region. It may be said as a final point that tropical forestry, at least as a beginning, has developed successful reconstruction processes and in part has already practically tested same.

This type of process depends upon the forest structure at hand, and has as aim a change to the function that the local population require of the reconstructed forest. Structure and function of South American forests were not the leading themes of my address, but the focus of all thoughtful forestry studies in South America.

Summary

South America is not only the richest forest region in the world (approx. 1 bill. ha), it also has the largest proportion of tropical forests with nearly 800 mill. ha. The present study deals mainly with the moist tropical evergreen and tropophytic forests of the northern part of the South American sub-continent. Characteristic is the abundance of tree species (40-80 species per ha), its intensive mixture quocient of mixture 1:5 to approx. 1:10), but varying depending to locality (low frequency of the majority of species). The dominances = the basal area, is between 15-60 m^2/ha. They have their minimum in the moist, tropophytic forests, maximum in the Andean cloud forest. The diameter distribution has a curve similar to that of an selection forest, which indicates that the permanence of these tropical forests is insured by continuous regeneration. Little knowledge is available as yet concerning the mechanisms of natural regeneration. There is evidence that the replacement process is very slow from individual to individual, in other words, regeneration is almost inapparent. Thus the forest as such can be considered, even in very small areas, as not time-limited. Only the individual tree has a beginning and an end.

The structure of natural forests in the South American tropical belt ensures a variety of protective functions in a satisfactory and sometimes even excellent way. For the same reason, however, such forests have only a limited economic productivity. Normally the timber production in quality and quantity is not satisfactory. Moreover, recreational functions are also lacking. This inability of the forests to meet a variety of needs, mainly resulting from structural weaknesses, puts the forester in a very difficult position. Thus their most important task is to convert the low-yielding natural stands into more economic stands based on a sustained yield wherever the conservation of forests has been recognized as necessary without reducing the important protective and social functions.

Resumen

Südamérica no sólo es la región más rica del mundo en bosques (aprox. 1000 mill. de ha), sino que posee además la mayor proporción de bosques tropicales con cercanamente 800 mill. de ha. El tema está circunscrito a los bosques higrofíticos del trópico, especialmente los siempreverdes y los tropófitos, de la parte septentrional de la América del Sur. Característico para ellos es la abundancia en especies arbóreas (40-80 por ha), su intensa mezcla (cuociente de mezcla 1:5 hasta aprox. 1:10), pero que varía fuertemente de acuerdo a la localidad (baja frecuencia de la mayoría de las especies). Las dominancias = áreas basales, oscilan entre los 15 y 60 m²/ha. Ellas son mínimas en los bosques tropófitos, máximas en el bosque nublado andino. La distribución de los diámetros es una curva decreciente típica de bosques de entresaca, lo que demuestra que la existencia de estos bosques está asegurada por una constante autorenovación. Pero los mecanismos de la regeneración natural son poco conocidos. Es evidente que por regla general el reemplazo se produzca paulatinamente pié a pié, es decir, la renovación de los rodales transcurre en períodos de tiempo muy largos, de una manera casi inaparente. El bosque como tal puede considerarse, incluso en una superficie bastante pequeña, como ajeno al tiempo. Un comienzo y un fin sólo existe para el árbol individual.

Las propiedades estructurales de los bosques naturales en la zona tropical sudamericana les confieren la posibilidad de realizar en forma satisfactoria hasta sobresaliente, variadas funciones protectoras. Sin embargo, por la misma razón no están en la situación, o sólo parcialmente, de hacerse cargo de una función productora. La producción de madera normalmente no es satisfactoria, tanto en calidad como en cantidad. Además faltan las condiciones primordiales para cumplir con funciones de recreación. La por la estructura condicionada incapacidad de los bosques sudamericanos para cumplir equilibradamente con las funciones requeridas le plantea a los forestales problemas extraordinariamente difíciles. Su tarea principal consiste — allá donde la conservación del bosque ha sido reconocida como imprescindiblemente necesaria — en transformar el bosque autóctono de baja producción en rodales económicos de rendimiento sostenido, sin tener que mermar por ello importantes funciones protectoras y sociales.

REFERENCES

AUBREVILLE, A. 1971. Quelques reflexions sur les abus auxquels peuvent conduire les formules d'evapotranspiration réelle ou potentielle, en matière de sylviculture et de bioclimatologie tropicale. Bois et Forêts des Tropiques, 136.

BERNAL, J. 1967. Estudio ecológico del bosque Caimital. *Rev. For. Ven.* 10 (15).

BRÜNIG, E. 1971. Forstliche Produktionslehre. Europäische Hochschulschriften. Bern - Frankfurt.

CAIN, St. *et al.* 1956. Application of some phytosociological techniques to brazilian rain forest. *American Journal of Botany* 43 (10).

FAO, 1971. Yearbook of Forest Products 1969-70. Rome.

FAZ, 1973. Entlaubungsmittel für Südamerika? Frankfurter Allgemeine Zeitung. *Natur und Wissenschaft* 89 (14) April.

FINOL, H. 1969. Posibilidades de manejo silvicultural para las reservas forestales de la región occidental. *Rev. For. Ven.* 12 (17).

FÖRSTER, M. 1973. Strukturanalyse eines tropischen Regenwaldes in Kolumbien. *Allg. Forst- und Jagdzeitung* (144) (1).

GOUJON, P. 1968. Conservation des dols en Afrique et a Madagascar. *Bois et Forêts des Tropiques*, 118.

HUECK, K. 1966. Die Wälder Südamerikas. G. Fischer, Stuttgart.

KLINGE, H. & W.A. RODRIGUES 1968. Litter production in an area of Amazonian Terra Firme Forest. Part. I. Litter-fall, Organic carbon and total nitrogen contents of litter. *Amazoniana* 1(4).

KNIGGE, W. 1969. Die Nutzung der tropischen Wälder Südamerikas. Deutsche Stiftung für Entwicklungsländer, DOK 440 C, S 1/69 ex.

KUMMEROW, J. 1966. Aporte al conocimiento de las condiciones climáticas del bosque de Fray Jorge. Bol. Tecn. No. 24 Fac. de Agronomía, Universidad de Chile.

LAMPRECHT, H. 1958. Der Gebirgs-Nebelwald der venezolanischen Anden. *Schweiz. Zeitschr. f. Forstw.* 4.

LAMPRECHT, H. 1969a. Einige waldbau-ökologische Überlegungen aus überregionaler Sicht. *Schweiz. Zeitschr. f. Forstw.* 1.

LAMPRECHT, H. 1969b. Über Strukturanalysen im Tropenwald. Festschrift Hans Leibundgut. Beih. Nr. 46 zu den Zeitschriften des Schweiz. Forstvereins.

LAMPRECHT, H. 1972. Einige Strukturmerkmale natürlicher Tropenwaldtypen und ihre waldbauliche Bedeutung. *Fortwissenschaftl. Centralbl.* 91 (4).

MAC 1961. Atlas forestal de Venezuela. Caracas.

PETRICEKS, J. 1959. Relación entre el área de bosques e intensidad de la agricultura migratoria en Venezuela. Bol. 4 IFLA, Mérida.

REHM, S. 1973. Landwirtschaftliche Produktivität in regenreichen Tropenländern 73 (2).

SUAREZ DE CASTRO, F. & A. RODRIGUEZ-GRANDAS 1962. Investigaciones sobre la erosion y la conservación de los suelos en Colombia.

VARESCHI, V. 1968. Comparación entre selvas neotropicales y paleotropicales en base a su espectro de biotipos. *Acta Bot. Venezuelica* 3: 1-4.

VEILLON, J.P. (s.a.). Estudio de la masa forestal de los bosques de las zonas bajas de Venezuela en relación con el factor climático: humedad pluvial y ensayo de la representación gráfica y matemática de las correlaciones. ULA Mérida.

WEIDELT, H.-J. 1968. Der Brandhackbau in Brasilien und seine Auswirkungen auf die Waldvegetation. Diss. Forstl. Fak. Uni. Göttingen.

Address of Author:

Prof. Dr. H. LAMPRECHT, Institut für Waldbau II, Büsgenweg 1, D-34 Göttingen, Germany.

APPENDIX

Examples of structure analysis for 3 forest types in tropical South America
(brief report)

1. *Evergreen rain forest.* Carare-Opón, Magdalena valley, Columbia,
100 - 300 m. u.s.l. (after FÖRSTER 1973)
climate   rainfall 2840 mm p.a., temperature 27.7°C
probe   1 hectare, from BHD 9.5 cm
results   trees = 571, species = 85, basal area = 28.1 m^2, mixture quotient ∼ 1:8
46.6% of the trees belonged to the 8 commonests species
49.8% of the basal area was occupied by the 8 dominant species
frequency classes I = 38, II = 26, III = 8, IV = 7, V = 6 species

vertical structure

canopy layer	upper	middle	lower
trees	52	272	247
species	24	66	57
mixture quotient	∼1:2	∼1:4	∼1:4

diameter distribution

10-20 cm	389 trees;	51-60 cm	9 trees;
21-30 cm	99 trees;	61-70 cm	7 trees;
31-40 cm	43 trees;	71-80 cm	6 trees;
41-50 cm	15 trees;	81- cm	3 trees.

2. *Evergreen cloudforest.* La Carbonera, Andes mountain, Edo. Mérida, Venezuela. 2300 m. u.s.l.
(after TORRES 1965)
climate   rainfall about 2000 mm p.a., temperature about 13°C
probe   1 hectare, from BHD 9.5 cm
results   trees = 1031, species 41, basal area = 46.2 m^2, mixture quotient ∼ 1:25
58.3% of the trees belonged to the 8 commonest species
66.1% of the basal area was occupied by 8 dominant species.
frequency classes I = 7, II = 7, III = 7, IV = 10, V = 10 species

vertical structure

canopy layer	upper	middle	lower
trees	132	199	622
species	29	29	37
mixture quotient	∼1:5	∼1:7	∼1:17

diameter distribution

10-20 cm	678 trees;	71- 80 cm	4 trees;
21-30 cm	198 trees;	81- 90 cm	6 trees;
31-40 cm	91 trees;	91-100 cm	0 trees;
41-50 cm	25 trees;	101-110 cm	2 trees;
51-60 cm	23 trees;	111-120 cm	0 trees;
61-70 cm	8 trees;	121-130 cm	1 tree.

3. *Raingreen moist forest.* El Caimital, Llanos occidentales, Venezuela 170 m. u.s.l. (after
LIZANO 1966)
climate   rainfall 1500 mm p.a., dry period December-April, temperature 26.5°C
probe   1 hectare, from BHD = 9.5 cm
results   trees 343, species 55, basal area 28.73 m^2, MQ ~1:6
42.6% of the trees belonged to the 8 commonest species
53.2% of the basal area was occupied by 8 dominant species
frequency classes I = 29, II = 11, III = 6, IV = 8, V = 1 species

vertical structure

canopy layer	upper	middle	lower
trees	35	85	223
species	16	34	46
mixture quotient	~1:2	~1:2.5	~1:5

diameter distribution

10-20 cm	190 trees;	81-90 cm	2 trees;
21-30 cm	60 trees;	91-100 cm	0 trees;
31-40 cm	39 trees;	101-110 cm	0 trees;
41-50 cm	21 trees;	111-120 cm	2 trees;
51-60 cm	13 trees;	121-130 cm	1 tree;
61-70 cm	7 trees;	131-140 cm	1 tree;
71-80 cm	6 trees;	141-160 cm	1 tree.

INUNDATION – FOREST TYPES IN THE VICINITY OF MANAUS

ULRICH IRMLER

From: Cooperation between Max-Planck-Institut für Limnologie, Abt. Tropenökologie, Plön (Holstein), and Instituto Nacional de Pesquisas da Amazônia (I.N.P.A.) Manaus – Amazonas, Brazil.

Abstract

In the vicinity of Manaus five types of inundation forest were differentiated according to the composition of the benthos community and the environmental conditions. The five types differ according to the influence of white and black water and the streaming or flow conditions. *Campsurus notatus* NEEDH. & MURPHY was found as typical for várzea forests with an inflow only of white water, *Eupera simoni* JOUSSEAUME in mixed water strongly influenced by white water and the Naïdidae in the lower courses of rain-forest streams.

1. Discussion of the terms inundation forest and igapó

1.1. Inundation forest

Fluctuations in the water level of the Amazon and its tributaries in the Manaus district are on a considerable scale, and cause these rivers each year to inundate wide areas of their valleys. According to water level measurements in Manaus harbour the average fluctuation was 10.10 m/year based on a 70 year period.[1] The maximum fluctuation was 14.13 m in 1909, the minimum 5.45 m in 1912.

The water level fluctuations, although showing marked variations in one year from the next, nevertheless have an exact annual periodicity. The highest water level at Manaus each year is in June; the lowest water level is recorded in the months October/November (fig 1). Forests near the river banks may be inundated from December to September/October; in years with a very high low-water even remaining inundated the whole year. The annual duration of inundation decreases with distance from the river bank, and the inundation forest changes gradually or abruptly, according to the form of the bank, to terra firme forest which is never inundated. A forest showing the described annual periodic inundation is here designated as inundation forest.

This forest type should be seen in relation to the types distinguished by ELLENBERG (1973): T 1.11 often-flooded floodplain forest, T 1.12 seldom flooded, T. 1.13 near ground-water. The difference according to these types lies in the very regular inundation of inundation forest, whilst the flood-plain forest can be flooded as the result of a heavy rainfall. That means inundation forest is associated with lowlands in the lower reaches of rivers which because of major climatic conditions in their source regions show large annual fluctuations in water level. Flood-plain forest, in contrast, is associated more with the upper reaches of rivers and streams, and the flooding is irregular, influenced by local weather conditions. In the vicinity of Manaus for example heavy storm-rain at the time of low water causes flooding of the flood-plain forest but not of the inundation forest.

These two distinctive conditions affect the composition of the fauna and flora.

---

1 Water level measurements provided by the Administração do Porto de Manaus (earlier Manaus Harbour), to whom thanks are due.

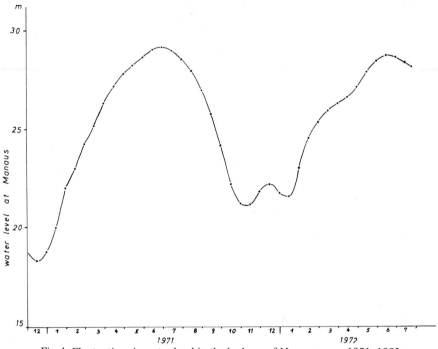

Fig. 1. Fluctuations in water level in the harbour of Manaus, years 1971, 1972.

In the inundation forest a marked annual periodicity in the fauna can be noticed. In both the terrestrial and the aquatic fauna extensive movements take place. Other survival mechanisms (e.g. dormant period) are also adopted by the terrestrial and the aquatic fauna of inundation forest (BECK 1969, 1972; SCHALLER 1969; IRMLER 1973; FITTKAU *et al.* 1975)

1.2. Igapó

Two further terms have led to confusion in the typification of inundation forest. They are várzea and igapó. Whilst the term várzea is used uniformly and correctly for the inundation lands of white-water streams (SIOLI 1956), igapó has been less narrowly employed.

The word igapó, the etymological origins of which are still not clear, is derived from the Indian Tupi language and means 'forest in standing water' or 'standing water in forest' (ROCQUE 1968, MOREIRA 1970). In vernacular usage therefore the term igapó can be applied to any forest in which there is for any reason standing water. There is then no distinction between the two previously mentioned types: the inundation forest and the floodplain forest.

Since many forest formations, otherwise easily distinguishable, are embraced by this original concept, the term itself has been variously employed in scientific literature. On one occasion the term igapó was used for both inundation forest and flood-plain forest, with amplification for other forest types (MOREIRA 1970). In many cases igapó was used simply for inundation forest but then in contrast to the

18

never-inundated terra firme forest. Both white-water forests and black-water forests have been described as igapó (GESSNER 1968). Other authors have used the term to indicate only the forests growing on the lowest terrain, inundated for almost the whole year (KATZER 1903, TAKEUCHI 1962). The obvious differences between inundation forests in the valleys of white-water streams and those of black-water streams has also been brought into the terminology with várzea forest for the first and igapó for the second (SIOLI 1951). This last definition has been more or less adopted in the more recent scientific literature.

2. Biocenotic aspects

2.1. Biocenotic arrangements of the samples

The existence of regular transitional forms between these various inundation forests has already led to the development of new terms (BECK 1971), and makes it possible, with the usage of faunistic samples from various points, to give a biocenotic classification of the inundation forests. In the years 1971/72 the soil macrofauna was investigated at 21 sampling stations in the vicinity of Manaus, each more or less at the time of high water (fig 2).[1] At each sampling station two or three samples were taken with a Birge-Ekman bottom sampler, when possible at a higher and at a lower site. Since this method can give no more than a very general view of the biotopes present, the sampling can only distinguish strongly contrasting types of forest. With a more exact examination of the biotopes a more refined division should result.

To present a classification from a biocenotic view-point, animals were selected that were found frequently in many of the samples, thus showing a relatively high presence value. For these animals individual or group dominance was calculated, that is the percentage of the selected species or group according to total abundance of the sample.

Five relatively separate groups can be differentiated (table 1). The first group in pure white water on an island in the Rio Solimões was investigated at only one sampling point, since intensive agricultural usage of the várzea in this district near Manaus had virtually destroyed the biotope. Here at the time of high water the mayfly *Campsurus notatus* NEEDH.& MURPHY (Ephemeroptera, Campsuridae) was dominant.

The second group is distinguished by the high constancy of *Eupera simoni* JOUSSEAUME (Bivalvia, Sphaeriidae). In addition other animals such as Caenidae (Ephemeroptera) and Ancylidae (Gastropoda) are relatively well represented whilst Naïdidae (Oligochaeta) are present only in small numbers. At these biotopes mixed water between black or acid clear-water and white water is present. A strong inflow of white water, however, occurs, as can also be seen from the transparency.

With the third group, the biotopes show an absence of *Eupera simoni*, and a high abundance of Naïdidae. Such biotopes resemble in their biocenotic composition the inundation forests of the following group in black or acid clear

1 This stay was made possible by Prof. H. SIOLI, Max-Planck-Institut für Limnologie, Abt. Tropenökologie, Plön, and Dr. PAULO DE ALMEIDA MACHADO, Instituto Nacional de Pesquisas da Amazõnia, Manaus-Amazonas, Brazil, to whom thanks are due.

19

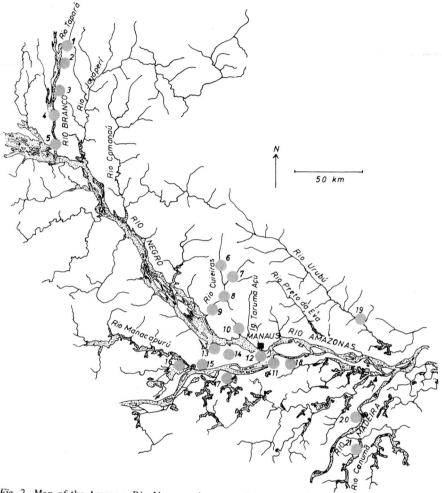

Fig. 2. Map of the Amazon, Rio Negro region around Manaus with the sampling sites. 1. Rio Tapará, 2. Ig. Agua Bôa, 3. Lg. Curumij, 4. and 5. islands in Rio Branco, 6. Rio Branquinho, 7. middle reach of Rio Cuieiras, 8. and 9. lower reach of Rio Cuieiras, 10. Rio Tarumã Mirím, 11. Ilha de Curarí, 12. Lg. Janauari, 13. P. Paricatuba, 14. Lg. do Limnão, 15. Lg. Calado, 16. Lg. Cabaliana, 17. Lg. Buiuçú, 18. Lg. do Chato, 19. Rio Urubú, 20. Lg. Bonfim, 21. Rio Canumã.

water, but have a relatively high transparency and pH. They contain mixed water, but the inflow of white water is feeble. In general they are also in further distance from the white-water rivers.

The fourth group includes the biotopes present in the lower courses of rain-forest streams. They show a high constancy and dominance of Naïdidae. As the example Igarapé Agua Bôa shows, the lower courses of clear-water streams with acid nutrient-poor water also seem to belong to this group. These forests of the lower courses with their slow moving, almost standing water during the high water period should be compared with the biotopes of the fifth group which are

20

Table 1. Biocenotic classification of inundation forests in the vicinity of Manaus.

locality	dominance (%) Campsurus notatus	Eupera	Caenidae	Ancylidae	Naididae	environmental factors µS$_{20}$/cm	pH	transparency m	water type
Ilha de Curari						46.6	6.1	0.4	white water
Lg. Janauari						49.0	6.1	1.05	
Lg. do Chato						39.0	5.8	1.3	
Lg. Curumij						12.3	5.8	0.5	mixed water with strong inflow of white water
Lg. Bonfim						32.9	6.0	1.5	
Lg. Cabaliana						33.9	6.4	2.2	
Rio Branco						14.9	6.0	0.7	
Ig. Água Bôa						14.2	5.7	1.5	
Lg. Buiuçú						38.8	6.7	2.0	mixed water with weak inflow of white water
Lg. do Limão						48.2	6.3	1.0	
Lg. Calado						27.8	6.0	2.1	
Rio Tarumã						7.1	5.0	1.1	weak flowing black or acid clear water
Ig. Água Bôa						6.2	4.6	5.0	
Rio Canumã						6.2	5.9	2.7	
P. Paricatuba						8.3	4.4	1.5	
Rio Urubú						9.9	4.2	1.4	
Rio Tapará						6.6	4.7	2.0	relatively strong flowing black or acid clear water
Rio Cuieiras						6.4	4.7	2.0	
Rio Cuieiras						7.2	4.6	2.3	
Rio Branquinho						7.2	4.3	2.0	

Legend: ▓ > 30% , ▦ 20-30% , ▒ 10-20% , ||||| < 10%.

21

associated with swiftly moving water. In these last forest formations the Naïdidae are only sporadically present.

2.2. Mixed water

In the course of the investigations it was clearly apparent that *Eupera simoni* is a characteristic species for biotopes of the second group, mixed water with a strong white-water inflow. Since the white water carries a high sediment load with a transparency of 0.3 to 0.5 m at high water, a lesser inflow of white water with an increased mixing of black or acid clear water results a higher transparency. It is now possible to compare the occurrence of *Eupera simoni* in terms of individual dominance at different sample areas with the corresponding transparency measurements (fig 3). This shows that *Eupera simoni* has an optimum range in the biotopes where a transparency of 1.0 to 1.2 m is present at high water. A decreasing transparency corresponding with an increased inflow of white water and an increased sediment load has a sharp effect on the abundance of *Eupera simoni*; similarly an increased influence of black or acid clear water.

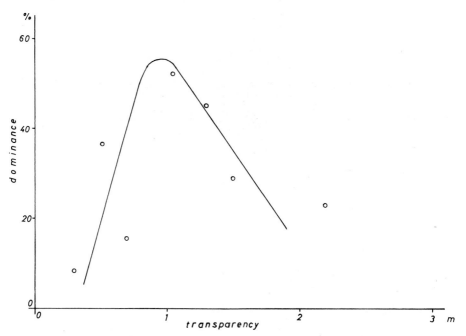

Fig. 3. Distribution of individual dominance of *Eupera simoni* in inundation forests of white and mixed waters with different transparency.

In a tributary of the lower Rio Branco, the Igarapé Agua Bôa, for comparative purposes three successive samples were made from acid clear-water to mixed water (table 2). *Eupera simoni* occurs first sporadically in the sample with the greatest white-water content. The clear water and the mixed water with a strong inflow of clear water are distinguished by a high dominance value of Naïdidae. Also present in clear-water are Chaoboridae (Diptera). Further differences are shown by the

22

	abundance /m²	dominance (%)			number of families	depth (m)	transparency (m)	O_2 in deep water (mg/l)	pH	μS_{20}/cm
		Naïdidae	Chaoboridae	Eupera						
a	45	50	50	–	2	5	5	0.5	4.6	6.2
b	200	55	–	–	4	3	1.8	0.6	5.5	11.9
c	750	3	–	3	9	2	1.5	1.2	5.6	14.3

Table 2. Biocenotic characteristics and environmental factors in the Igarapé Agua Bôa, a. clear water, b. mixed water with a strong inflow of clear water, c. mixed water with a weak inflow of clear water.

number of families, increasing form clear water to mixed water with a stronger inflow of white water. Environmental factors in the same sequence are increasing O_2, pH and conductivity values, whilst the transparency is greatest in clear water. As already shown by table 1 the mixed water biotopes of the biocenosis with strong inflow of black or acid clear-water are also characterized by the fauna of black or acid clear waters.

2.3 Rain-forest streams

The smaller rain-forest streams are dammed by the high water of the main rivers as far back as the sedimentation zone or middle course (FITTKAU 1967). Due to FITTKAU (1967) the inundation forests of the rain-forest streams in the vicinity of Manaus can be divided into those of the lower course and those of the middle course. In table 3 samples in various inundation-forest sites laying nearest to the river are compared. The lower-course biotope shows again the same faunal poverty and the same environmetal conditions as was expressed in the previous tables.

sampling station		abundance /m²	number of families	dominant family	μS_{20}/cm	pH	transparency (m)	O_2 (mg/l) in deep water	current cm/sec	depth (m)
Rio Cuieiras	I	850	4	Chirono.	5.6	4.8				0.5
	II	620	3	Chirono.	6.4	4.7	2.0	5.1		2
	III	530	4	Chirono.	7.2	4.6	23			4
Rio Branquinho		450	5	Cerato.	8.6	4.3	20	7.1	16	2
Rio Canumã		45	1	Chaobo.	6.2	5.9	2.7			7
Rio Tapará		90	2	Chaobo.	6.6	4.6	20	0.4		6
Ig. Água Bôa		45	2	Naid./Cha.	6.1	4.6	5.0	0.5		5
Rio Cuieiras	IV	–	–	–	7.2	4.6	3.0		0	8
	V	–	–	–	7.2	4.6	2.7	0.4	0	8

Table 3. Biocenotic characteristics and environmental factors of that part of inundation forests nearest to the open rain-forest streams in the vicinity of Manaus.

The inundation forests of the middle courses show a richer fauna. Both the abundance and the number of families are higher than in the lower courses. In the lower courses there are above all Naïdidae and Chaoboridae, but here there are particularly the Chironomidae (Diptera) that comprise the major part of the fauna.

Due to the greater distance from the main river, a lower inundation height occurs in this biotope. The higher concentration of oxygen may be due to the greater velocity of current in these reaches of the rivers. At a site of the Rio Branquinho on 24.6.72 this was 16 cm/sec. Both the higher velocity of current and the higher oxygen content seem to constitute the essential difference compared with the lower course inundation forest, and lead to another form of zoocenosis. A differentiation of inundation-forest types on the rain-forest rivers with black or acid clear water can thus be affected.

The ecological conditions in the middle course of a rain-forest stream can be more exactly demonstrated due to a transect in this reach of the Rio Branquinho. This investigation was made in 24.6.72 during high-water. At three sites three samples were taken respectively: one site was nearest to the open river (table 4 a), the second in the centre (b) and the third near the terra firme (c).

sampling station	abundance / m²	number of families	dominance (%) (abundance/m²)			O_2 (mg/l)	pH	μS_{20}/cm	depth (m)
			Cerato-pogonidae	Epheme-roptera	Chirono-midae				
a	450	5	52 (222)	17 (74)	20 (90)	7.1	4.3	8.6	2
b	490	5	9 (45)	3 (15)	73 (356)	6.2	4.3	8.3	1.5
c	120	4	25 (30)	12 (15)	50 (60)	6.2	4.3	8.0	0.5

Table 4. Biocenotic characteristics and environmental factors in a transect of the middle reaches of the Rio Branquinho. a. near the open stream, b. central part, c. near the terra firme.

The distribution of abundance shows highest values in the sites near the river and the terra firme. The distribution of different taxa also varies in this transect. In the site near the open river there are particularly the Ceratopogonidae (Diptera) and the Ephemeroptera that are most numerous whilst the Chironomidae reach higher abundance in the central part and in the site near the terra firme.

The oxygen content, which decreases from the open river to the site near the terra firme, suggest that the velocity of current is an important environmental factor due to the ecological structure of this biotope. At the site nearest to the open river, the current removes the litter so that a sandy bottom developes, whilst in the parts nearest to the terra firme a thick litter layer leads to higher decomposition and oxygen consumption.

3. Vegetation arrangement

Due to the floristic composition of várzea and igapó only a few studies are published (AUBREVILLE 1961, BOUILLENNE 1930, DUCKE & BLACK 1954, GESSNER 1968, HUBER 1909, HUECK 1966, TAKEUCHI 1962). Such descriptions consist generally only of a list of species without either exact localities or the ecological conditions of the site. Only HUBER (1909) differentiated various vegetation types, while the other authors used the general term 'várzea' for a variety of ecologically different areas. It is therefore not possible to compare the biocenotical classification presented with corresponding plant communities of inundation forests.

Most of the plants noted by authors are only mentioned once so that it is impossible to decide whether the plants are rare and present only at definite sites in the várzea or igapó, or if they are in fact more widely distributed and not recognized by other authors. Table 5 shows the few species mentioned by more than one author, or by one author and additionally found in an inundation forest investigated in this study. A difference between várzea and igapó obviously exists. Descriptions of igapó vegetation are published only by GESSNER, HUBER and HUECK. Since they do not give details of the water chemistry it is not possible to decide whether we are dealing with pure black-water, or mixed water with a strong black-water inflow.

Species	Várzea	Igapó
Ceiba pentandra	H/DB/A/G/B/HUECK	–
Olmediophaena maxima	H/DB/G/B/HUECK	–
Bombax munguba	H/DB/G/B/HUECK	–
Sterculia elata	DB/G/B/HUECK	–
Cecropia spec.	H/DB/G/B/HUECK	–
Calycophyllum spruceanum	H/DB/G/B/HUECK	–
Hura crepitans	A/H	–
Pithecolobium niopoides	DB/HUECK	–
Virola surinamensis	DB/HUECK	–
Couralia taxoñora	G/H	–
Carapa guianensis	DB/HUECK	–
Couroupita subsessilis	H/HUECK	–
Piranhea trifoliata	DB	–
Gustavia augusta	H	–
Vitex cymosa	H	–
Swartzia	–	G/H/HUECK
Eugenia inundata	–	G/HUECK
Macrolobium acaciaefolium	–	H/HUECK
Campsiandra laurifolia	–	G/H

Table 5. The plants of the várzea and igapó mentioned by HUBER (H) DUKE & BLACK (DB) AUBREVILLE (A) GESSNER (G) BOUILLENNE (B) and HUECK. Only those plants are included that are mentioned by more than one author, or if mentioned by only one author they were additionally found in the areas investigated in this study.

It seems to me that *Bombax munguba* MART. and *Vitex cymosa* BERT. are characteristic for white water areas in the vicinity of Manaus, although *Vitex cymosa* also occurs in mixed water areas with a strong white water inflow. In the black and acid clear-water *Jugastrum coriaceum* MIERS, and *Macrolobium acaciaefolium* seem to be important. HUBER (1909) mentioned *Hevea spruceana* MUELL. to be characteristic for mixed water areas.

4. Schematic classification

Transitional forests between the inundation forests of the middle and lower

25

reaches of rain-forest streams exist as well as those between the mixed water, and the black or acid clear water, which could not all be taken into account for the schematic representation. This schematic representation can therefore, not be regarded as an exact reproduction of the natural conditions.

According to choose terms for the various inundation-forest types it was attempted to tale into account both the general scientific and the vernacular usage in Amazonia. This schematic representation should therefore serve to explain the various concepts connected with these terms in order to achieve in future a more uniform terminology.

inundation forest					*flood-plain forest*
seasonally inundated forest (on plains of large rivers)					*flooding non-periodic (on small streams)*
várzea forest on white water rivers			*igapó* on rain-forest rivers with black or acid clear water		
inflow of white water			current		
pure	*strong*	*little*	*weak*	*relatively strong*	
low transparency	moderate transparency	high transparency	high transparency	high transparency	
relatively high conductivity	relatively high conductivity	relatively high to moderate conductivity	low conductivity	low conductivity	
pH: 6-7	pH: 6-7	pH: 5-7	pH: 4-6	pH: 4-5	
dominant group: Campsurus	dominant group: Eupera	dominant group: Naïdidae	dominant group: Naïdididae	dominant group: Chironomidae	

Table 6. Outline of characteristics of the discussed inundation forest types.

Table 6 represents the schematically different inundation-forest types. The term inundation forest should be used supplementarily to the terms used by ELLENBERG (1973) like flood-plain forest, lowland rain-forest etc. The exact term for the biotopes investigated is therefore 'inundation forests of the warm humid evergreen broadleaved forest region'. These consist, in the vicinity of Manaus, of várzea forests and igapós. The term várzea forest was used for all inundation forests with white-water or white-water inflow.

It was possible to differentiate broadly still further: forests with pure white-water with relatively high sedimentation, and a fauna distinguished by the presence of *Campsurus notatus* and in lesser abundance *Eupera simoni*; mixed-water forests with a strong inflow of white water, still moderate sedimentation and a high abundance of *Eupera simoni*; and mixed-water forests with a weak white-water inflow forming the transition to inundation forests of black or acid clear water designated as igapós. The igapós could be further subdivided into two types, that with almost standing water, and that with swiftly moving water. Between the latter and the various types of flood-plain forest numerous transitional forms may be expected.

The inundation forests are arranged in the Amazonian Landschaft in a definite situation in relation to the Amazon (fig 4). A schematic Landschaft based on

26

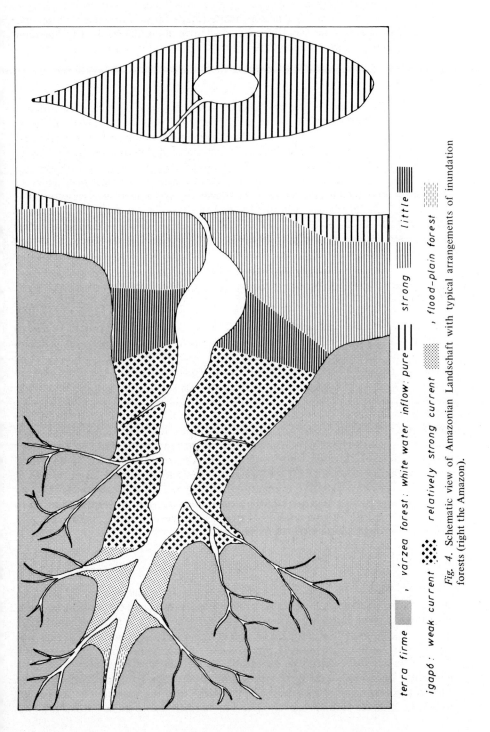

terra firme ▨ , várzea forest: white water inflow: pure ☰ strong ☰ little ☷

igapó: weak current ⣿ relatively strong current ▨ , flood-plain forest ░

Fig. 4. Schematic view of Amazonian Landschaft with typical arrangements of inundation forests (right the Amazon).

inundation forests along an Amazonian river shows the igapó with relatively strong current as furthest from the Amazon, while várzea forest with a pure white water inflow is situated on an island or on a small strip along the bank. The remaining types are arranged in succession between these extreme forest formations.

Zusammenfassung

Nach einer Erklärung der Begriffe 'Überschwemmungswald' und 'Igapó' wurden an Hand von 21 Probenstellen in der Umgebung von Manaus nach biozönotischen Gesichtspunkten fünf verschiedene Überschwemmungswaldtypen aufgestellt. Für die drei vom Weißwasser beeinflußten Typen wurde die Bezeichnung Várzeawald, für die zwei an den Schwarz- oder Klarwasser führenden Regenwaldflüssen der Begriff Igapó gebraucht. Alle Typen sind durch zahlreiche Übergänge miteinander verbunden und bilden eine sukzessive Anordnung von reinen Weißwasserwäldern bis hin zu den Igapòs der Mittelläufe amazonischer Regenwaldflüsse. *Campsurus notatus* NEEDH. & MURPHY wurde als typisch für Várzeawälder mit reinem Weißwasser, *Eupera simoni* für Mischwässer mit starkem Weißwassereinfluß und Naïdidae für die Unterläufe der Regenwaldflüsse in der Umgebung von Manaus gefunden.

Summary

The terms 'inundation forest' and 'igapó' were explained. Based on 21 sampling stations in the vicinity of Manaus the benthos communities of different inundation-forest types were studied in the years 1971, 1972. The inundation forest of the várzea was called várzea forest. For forests of the rain-forest streams with black or acid clear water the term igapó was used. The benthos community and the environmental factor study distinguished five inundation-forest types. Whereas in the three types of the várzea forest sedimentation is important, in the rain-forest streams the streaming or flow conditions differentiate two forest types. *Campsurus notatus* NEEDH. & MURPHY was found particularly in the várzea forest with an inflow of white water, *Eupera simoni* JOUSSEAUME in mixed water strongly influenced by white water and the Naïdidae in the lower courses of rain-forest streams. The five inundation-forest types are successively distributed along the river sides from the várzea forest influenced only by white water up to the middle courses of the rain-forest streams. Between neighbouring community types transitional communities were detected.

Resumo

Os termos 'floresta inundável' e 'igapó' são explicados. Com base em 21 estações de coleta na vizinhança de Manaus foram estudadas as comunidades bentônicas de diferentes tipos de floresta inundável no ano de 1971, 1972. A floresta inundável da várzea foi denominada floresta de várzea. O termo igapó foi isado para aquelas dos rios de floresta pluvial com água preta ou clara ácida. Uma divisão em cinco tipos de floresta inundável resultou do estudo da comunidade bentônica e do fator ambiente. Enquanto a sedimentação é importante nos três tipos de floresta de várzea, nos rios de floresta pluvial as condições de correnteza diferenciam os dois tipos de floresta inundável. *Campsurus notatus* NEEDH. & MURPHY foi encontrado particularmente na floresta de várzea com influxo apenas de água branca, *Eupera simoni* JOUSSEAUME em água mixta fortemente influenciada por água branca, e os Naïdidae nos coursos inferiores dos rios de mata pluvial. Os cinco tipos de floresta inundável estão distribuidos ao longo das margens dos rios, desde a floresta de várzea influenciada apenas por água branca até os cursos médios dos rios de mata pluvial. Comunidades transitórias foram encontradas entre tipos de comunidade vizinhas.

REFERENCES

AUBREVILLE, A. (1961): Etude écologique des principales formations végétales du Brésil. – Centre Tecn. For. Trop., Nogent-sur Marne: 1-268.
BECK, L. (1969): Zum jahreszeitlichen Massenwechsel zweier Oribatidenarten (Acari) im neotropischen Überschwemmungswald. – *Verh. dtsch. Zool. Ges. Innsbruck*: 535-540.
BECK, L. (1971): Bodenzoologische Gliederung und Charakterisierung des amazonischen Regenwaldes. – *Amazonia* 3: 69-132.
BECK, L. (1972): Der Einfluß der jahresperiodischen Überflutungen auf den Massenwechsel der Bodenarthropoden im zentralamazonischen Regenwaldgebiet. – *Pedobiologia* 12: 133-148.
DUCKE A. & BLACK G.A. (1954). Notas sôbre a Fitogeografia da Amazônia Brasileira.– *Bol. Téc. Inst. Agr. do Norte* 29: 3-62.
ELLENBERG, H. (1973): Ökosystemforschung – Berlin, Heidelberg, New York.
FITTKAU, E.J. (1967): On the ecology of Amazonian rain-forest streams. – Atas do Simpósio sôbre a Biota Amazônica 3 (Limnologia): 97-108.
FITTKAU, E.J. IRMLER, U., JUNK, W.J., REISS, F. & SCHMIDT, G.W. (1975): Productivity, biomass, and population dynamics in Amazonian water bodies. In: Tropical ecological systems. F.B. GOLLEY & E. MEDINA (eds.). pp. 289-331. Springer, New York.
GESSNER, F. (1968): Zur ökologischen Problematik der Uberschwemmungswälder des Amazonas – *Int. Rev. ges. Hydrobiol.* 53: 525-547.
HUBER, J. (1909): Mattas e madeiras amazonicas. – Bolm. Mus. para. 'Emilio Goeldi', Belém-Pará 6: 91-225.
HUECK, K. (1966): Die Wälder Südamerikas. – Stuttgart.
IRMLER, U. (1973): Population-dynamic and physiological adaptation of *Pentacomia egregia* CHAUD. (Col. Cicindelidae) to the Amazonian inundation forest. – *Amazoniana* 4: 219-227.
KATZER, E. (1903): Grundzüge der Geologie des unteren Amazonasgebietes (des Staates Pará in Brasilien). Leipzig.
MOREIRA, E. (1970): Os Igapós e seu aproveitamento – Imprensa Universitária, Belém-Pará.
ROCQUE, C. (1968): Grande Enciclopédia da Amazônia. – Belém-Pará.
SIOLI, H. (1951): Zum Alterungsprozess von Flüssen und Flußtypen im Amazonasgebiet. – *Arch. f. Hydrobiol.* 45: 267-283.
SIOLI, H. (1956): Uber Natur und Mensch im brasilianischen Amazonasgebiet. – *Erdkunde* 10 (2): 89-109.
SCHALLER, F. (1969): Zur Frage des Formensehens bei Collembolen. – *Verh. dtsch. Zool. Ges. Innsbruck*: 368-375.
TAKEUCHI, M. (1962): The structure of the Amazonian vegetation VI Igapó. – *Journ. Faculty of Science, Tokyo, Sect.* 3, Bot. 8: 297-304.

Authors address
Dr. ULRICH IRMLER, Max-Planck-Institut für Limnologie,
Abt. Tropenökologie, D-232 Plön, B.R.D.

A CONTRIBUTION TO THE PHYTOGEOGRAPHICAL STUDY OF TEMPERATE CHILE

VICTOR QUINTANILLA

Introduction

South American Chile is a country of great latitudinal extent, and of a great variety of landscapes. It is over 4200 km from north to south, and 180 km from east to west. The surface area is about 742,000 km^2.

The great latitudinal extent (from 17°30' to 56°30') gives rise to a great range of climates. Thus in the north is an extensive zone of a most-marked desert climate. Southward in succession are semi-desert, Mediterranean and temperate climates, followed in the Patagonian region and in Tierra del Fuego by a cold climate. The numerous climates are accompanied by important differences in the vegetative cover, in the hydrographic regimes, in the way of life of the population and in exploitation of natural resources.

The presence of the Andean cordillera along the whole eastern border gives a particular character to the landscape, and the cordillera plays an important role in the climatic types. The altitude of the Chilean cordillera diminishes however from north to south. Thus in the tropical zone many peaks rise above 6000 m., whilst in the Patagonian zone and in Tierra del Fuego the summits rarely attain more than 2000 m.

Latitude and altitude have a very strong influence on the vegetation of Chile, its extent, distribution and composition. Local factors, such as soil, water, and exposure affect the composition of the forests to a certain point, and man can be held responsible for their actual state.

1. Temperate Chile: Geographical and climatic description

Temperate Chile begins in the transitional provinces of Arauco and Bio-Bio and extends as far as the province of Chiloé, a distance of 1000 km. The temperate vegetation zone covers nearly one sixth of the total surface area (125,000 km^2) and reaches from 37° S to 44° S. In general this part of Chile has a temperate climate with a dry summer in the north, and a markedly wet climate the whole year in the south. Nevertheless, with such a latitudinal extent there are major differences in pluviosity and in temperatures. Thus Temuco in the central plain has an average annual rainfall of 1350 mm, Osorno 1500 mm, Puerto Montt 1750 mm and Valdivia, close by the sea, 2500 mm. It is estimated that the western slopes of the Andes both, in the elevated areas, as well as in the islands, receive 3000 mm or even more, whilst the transandean zone in Chiloè province is extremely dry, receiving 200 to 400 mm per annum. To these sharp differences in rainfall correspond distinctive types of forest, particularly on the eastern and western slopes of the cordilleras. In general temperatures diminish from north to south, and from sea level to the

Fig. 1. Temperate Chile and its situation in S. America.

mountain summits. The limit of tree growth varies from about 2,000 m. a.s.l. in the north to about 1000 m. in the south.

This vast region contains the greater part of the forests and natural woodlands of Chile, about 60% of the total forested surface, and 80% of the natural productive forests of the country.

32

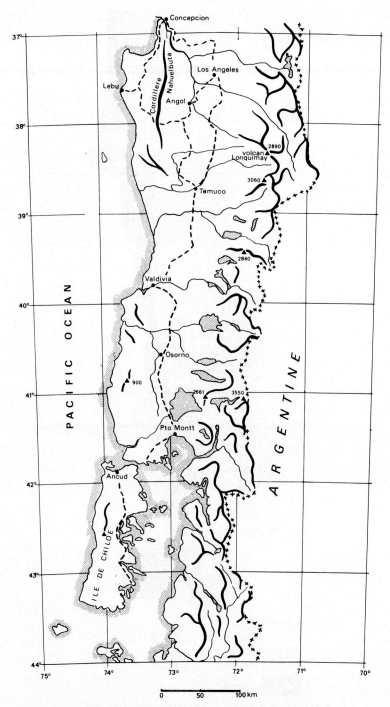

Fig. 2. The geographical zone of temperate Chile.

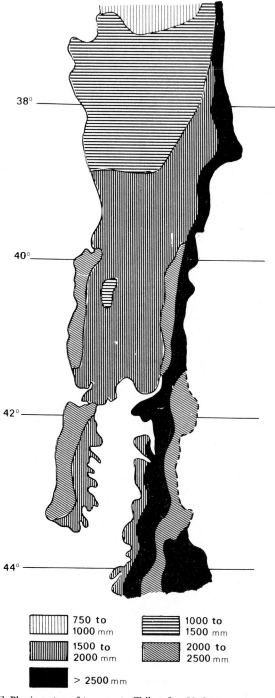

Fig. 3. Pluviometry of temperate Chile, after V. QUINTANILLA.

2. The forests

A. The distribution of forests

The north of the region is transitional in vegetation. One passes from a Mediterranean matorral and from *Acacia caven* to woodlands of a forest-park type with *Nothofagus obliqua* and *Boldea chilensis*.

The great central plain, contained between the two cordilleras, has today scarcely any natural vegetation because of intensive usage over a long period.

South of about the 39th parallel there is a certain homogeneity in the vegetation formations, principally because of the increase in rainfall. On the slopes of the cordilleras the dominant formations consist of different species of *Nothofagus* and of conifers.

More to the south, beyond the 42nd parallel, on the coastal cordillera, is a luxuriant forest, showing no seasonal pause, a mixture of *Nothofagus dombeyi* and *Nothofagus obliqua* with *Podocarpus* ssp., lianes, epiphytes and tree ferns, which is called: the Valdivian forest.

In the Andes the altitude accentuates the winter cold and there is no chrub layer. Deciduous species appear.

B. Description of the forests

In the temperate forests of Chile it is estimated that seven species of *Nothofagus*, two evergreen and five deciduous, constitute together two thirds of the population. The remaining third consists of a number of other broadleafed species, mainly with evergreen leaves, and eight conifers.

The mixed populations are much more numerous than the pure stands. Nevertheless there are exceptions, as for example the forests of coigue (*N. dombeyi*) an evergreen, in regions with a high rainfall. Other examples are of conifers auracaria, alerce and de las Guaitecas cypress (*Pilgerondendrum uviferum*) at the particular stations where these species grow. Given that the occurence of each species is limited by latitude, altitude and other factors there are at each station, even those with mixed woodland, rarely more than 10 species, of which three or four are predominant.

1. The broadleafed trees. deciduous species

Nothofagus spp. are found in the driest zones in the north of the region (provinces of Arauco, Malleco, Cautin) with the exception of the plateaux and exterior slopes of the coastal cordillera. Both the roble (*N. obliqua*) and the rauli (*N. procera*) are present. The first thrives at well drained stations of the central plain, in general up to altitudes of 900 m. although this varies with direction of exposure. At about 600 m. a mixture of *N. dombeyi* and *N. Procera* occurs. The rauli has an altitudinal range of 900 m. (between 300 and 1200 m.) in particular in the Andes. The provinces of Osorno and Lanquihue can be considered the southern limit of these species.

Two other deciduous species of Nothofagus are also present, lenga (*N. pumilio*) and ñirre (*N. antarctica*) but at higher altitudes, in general from 1300 m. to the altitudinal limit of tree growth, 2000 m. in the north. However, the ñirre is often

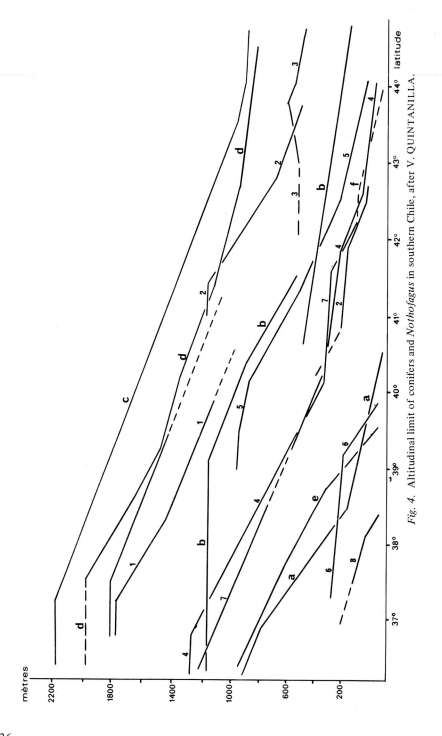

Fig. 4. Altitudinal limit of conifers and *Nothofagus* in southern Chile, after V. QUINTANILLA.

36

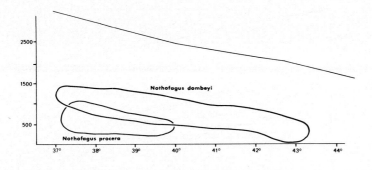

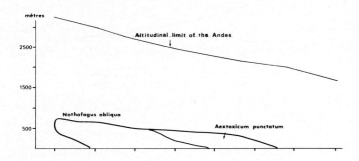

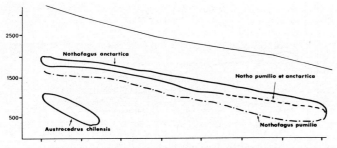

Fig. 5. Altitudinal distribution of *Nothofagus* in the Chilean Andes, after V.QUINTANILLA.

present as low woodland or scrub from 800 m. This limit is higher in the bottom of badly drained valleys.

The other deciduous species in this sector are often associated with evergreen species.

The evergreen species

The principal broadleafed species is the coigüe which tends to constitute pure stands beween 1000 and 1200 m. in the north of the region. It is very abundant forming more than 40% of the forests of the Valdivian province. Tepa (*Laurelia serrata*) is in quantity, the second most persistant tree in the formation, often

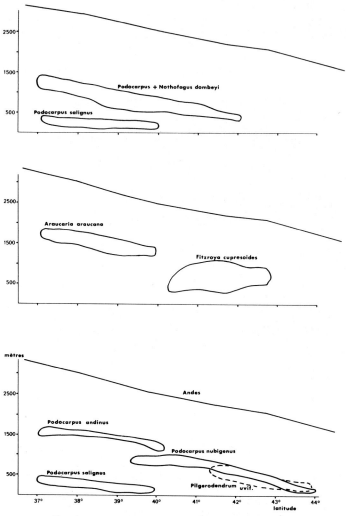

Fig. 6. Altitudinal distribution of Chilean conifers.

forming a distinct secondary stratum. Other evergreen trees are part of the preceding formations. The laurel (*Laurelia sempervirens*) in association with the roble is characteristic of certain parts of the central plain in the provinces of Osorno and Llanquihue: other species are ulmo (*Eucryphia cordifolia*), olivillo (*Aextoxicum punctatum*), tineo (*Weinmania trichosperma*) lingue (*Persea lingue*) and canelo (*Drymis winteri*)

In a lower stratum is found luma (*Amomyrthus luma*), tepu *Tepualia stipularis*, arrayan (*Myrceugenia apiculata*) avellano (*Gevuina avellana*) and ciruelillo (*Embothrium coccineum*). Most of these species are characteristic of moist stations along rivers or in marshy places where they form a low dense forest. Occasionally they are co-dominant.

38

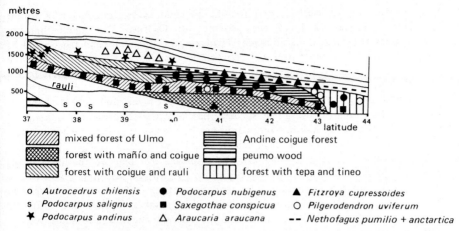

métres

mixed forest of Ulmo	Andine coigue forest
forest with mañío and coigue	peumo wood
forest with coigue and rauli	forest with tepa and tineo

o *Autrocedrus chilensis* ● *Podocarpus nubigenus* ▲ *Fitzroya cupressoides*
s *Podocarpus salignus* ■ *Saxegothae conspicua* ○ *Pilgerodendron uviferum*
✱ *Podocarpus andinus* △ *Araucaria araucana* -- *Nethofagus pumilio + anctartica*

Fig. 7. Altitudinal profile of vegetation in temperate Chile on the west slope of the Andes. Based on the work of J. SCHMITHUSEN.

The rest of the zone (the plateaux and the slopes of the coastal cordillera which run down to the sea in the north of the region, the central plains below the 41st parallel, all the islands and maritime slopes of the continent south of Puerto Montt) is covered by a typical forest of evergreens. This may be considered, in view of the heavy precipitation, a rain forest, or a dense temperate forest. Three persistant species of *Nothofagus* are characteristic: coigüe found as far as the south; the roble of Chiloé (*N. nitida*) which begins in the province of Osorno but is present particularly in Chiloé towards the south, and the roble of Magallanes (*N. betuloides*) beginning from the coastal Cordillera, from Valdiria, down to the south. The canelo and the luma accompany these species. All the other evergreen trees mentioned in association with the deciduous forests in the northern part of the region are found in the rain forest, but some disappear south of the Chiloé province.

2. The conifers

The association, often close, of conifers and broadleafed trees, despite the preponderance of the latter, is a characteristic trait of the natural forests of Chile. There are eight species of indigenous conifers.

Araucaria (*Araucaria araucana*) extends in the Andes between 37° and 39° 30's, from an altitude of 1200 m up to 1800 m, and equally in the area of Nahuelbuta on the coastal cordillera, thinly north of 38th parallel. It forms either pure populations, or is mixed with the coigüe and the lenga.

The alerce (*Fitzroya cuppressoides*) is a species of the rain forest. It is restricted in the coastal cordillera of Valdivia province to altitudes above 300 m. More to the south in the province of Llanquihue it is found a little in the central plain on marshy soils. It is also present in humid localities and in the Andino sectors of Llanquihue and Chiloé provinces. Where conditions are favourable it grows to a great size and may reach an age of 2000 or more.

Among the other conifers the cordillera cypress *Austrocedrus chilensis* the lleuque (*Podocarpus andinus*) and the mañio (*Podocarpus salignus*) are found particularly in the north of the province of Osorno. The mañio hembra (*Saxegothaea conspicua*), mañio macho (*Podocarpus nubigenus*) and Las Guaitecas cypress (*Pilgerodendrum uniferum*) characterise particularly the areas of heavy rainfall, and are found further to the south. The cordillera cypress prefers rocky sites. The lleuque is relatively rare. The mañio is present on both the cordillera as well as the central plain. It is usually a member of the underwood, beneath dominant deciduous or evergreen species. Las Guaitecas cypress is dominant at humid sites. The mañio hembra and mañio macho are usually subdominant species, though found in association with broadleafed trees.

The forest of southern Chile are characterised by their dense underwood, formed in parts by a mixture of ombrophilous species such as tepa, luma and the three mañios, but more generally by a dense vegetation made up of various species of the genus *Chusquea* (colihue and quila). This makes regeneration difficult, particularly as the species of greatest commercial value such as raulí and roble are light demanding.

At greater altitudes in the Andes the underwood is often constituted by hombrophyle species of canelo (*D. wynteri*).

The description of the natural forests of the temperate region of Chile shows that their distribution is irregular. At the present time the main areas of exploitation are the provinces of Valdivia, Cautin and Malleco. These provinces produce 60% of the country's timber. The standing crop of the productive forests has been estimated as 1500 millions m^3, and annual growth 17 millions m^3. The annual loss due to natural causes such as decay, wind action, insect pests, desease and fire is estimated at 43 million m^3.

These figures show a grave situation which, if not redressed, will lead to the loss of the natural productive forests in 50 to 60 years.

BIBLIOGRAPHY

BAUER, P. (1958: Waldbau in Sudchile. *Bonner Geogr. Abh.* 23. Bonn.

BRUGGEN,J.s(1948): La expansion del bosque en el sur de Chile en la época postglacial. Rvta. Universitaria. XXXIII. Santiago de Chile.

—. Corporacion Chilena-Clasificacion Preliminar del Bosque Nativo de la Madera Chile. (1967): Con dos mapas, 1:500.000. Instituo Foestal. Chantiago de Chile.

—. (1971): Maderas. Recopilacion preparada por H. Torres, Santiago de Chile.

DI CASTRI, FCO, (1968): Esquisse ecologique du Chile. Biologie de l'Amerique Australe. Volume IV. C.N.R.S. Paris.

ELIZALDE, M. RAFAEL. (1958): La sobrevivencia de Chile. Ministerio de Agricultura. Santiago de Chile.

FUENZALIDA, V. Hbto. (1965): Climatologia-Biogeografia. In, Geografia Economica de Chile. Texto refundido. CORFO Santiago de Chile.

HAUMANN, L. (1916): La fôret valdivienne et ses limites.

HUECK, K. (1966): Die Wälder Südamerikas. Gustav Fischer. 422 pp. Stuttgart.

LAUER, W. (1961): Wandlungen im Landschaftsbild des südchilenischen Seengebietes seit Ende der Spanischen Kolonialzeit. *Schriften des Geographischen Instituts der Univ. Kiel* 20: 227-276.

MUNOZ, P.C. (1966): Sinopsis de la Flora Chilena. Ediciones de la Universidad de Chile. Santiago de Chile.

OBERDORFER, E.s(1960): Pflanzensoziologische Studien in Chile, ein Vergleich mit Europa. 208 pag. Verlag von J. Cranier, Weinheim.

PISANO, E. (1950): Mapa de las formaciones vegetales de Chile. In: Geografia Economica de Chile. 1, Santiago de Chile.

—. (1954): La vegetacion de las distintas zonas geograficas chilenas. Rv a2 Geografica de Chile. Terra Australis. No ll.

—. (1966): Zonas Biogeográficas de Chile. In: Geografica Economica de Chile. Primer Apéndice. Santiago.

QUINTANILLA, V.G. (1974): Les formations végétales du Chili tempéré au 1:1.000.000.— Documents de Cartographie Ecologique, XIVI. La Biologie Végétale, Univ. de Grenoble, France.

REICHE, C. (1896-1911): Flora de Chile. 6 Volumes. Santiago de Chile.

SCHMITHÜSEN, J. (1956): Die raumliche Ordnung der chilenischen Vegetation. *Bonner Geogr. Abh.* 17: 1-89 Bonn.

—. (1960): Die Nadelhölzer der Waldgesellschaften der Sudlichen Anden. *Vegetatio*, 9. The Hague.

SCHWABE, G.H. (1952): Aspectos ecologicos de Chile. Re vta. Atenea. Universidad de Concepcion. Concepcion.

Author's address:

Prof. Dr. V. QUINTANILLA, Institut de Géographie, Université Catholique Valparaiso, Chile.

ÖKOSYSTEME OSTPARANAS

HANS JAKOBI

Abstract

Ecological research in tropical region is badly needed: old and valuable ecosystems are threaten-
ed to become destroyed by man before they are known by science. It is the first tentative to
classify the ecosystems of the State Paraná by ecodynamic approach. Rich in peculiar environments and striking changes of the main ecological factors make the research of special interest.
  The author analyses 17 natural ecosystems for Eastern Paraná: 1. The continental shelf, 2.
The thalassepsammal (beach communities), 3. The restingal and strip of dunes, 4. The littoral
forest (ephiepiphytal and phytothelmal), 5. The tholomigal (soft bay soils), 6. Island mangrovial, 7. Island forest, 8. Island psammal and psephal, 9. Mangrovial, 10. Pantanal (swamps), 11.
Bananal (man-made ecosystem), 12. Littoral tropical rainforest, 13. Cataractal (water falls),
14. Nanobambusial (small bamboo biotopes), 15. Subtropical rainforest, 16. The valley of
Iguaçú River, 17. The araucarial.
  A complex profile through eastern Paraná outlines these studied ecosystems stressing climatic, geomorphological and ecological particulares of terrestrial and aquatic biotopes.

Tropenökologische Forschungen werden von tropenheimischen Institutionen nur
in Ausnahmefällen betrieben, und unsere Kenntnisse über die funktionalen
Zusammenhänge stehen noch am Anfang. Diesen Tatsachen steht eine rasche
technologische Erschließung der Tropengürtel gegenüber, die zu einer radikalen
Veränderung der Naturlandschaften führt. Beschleunigte Bevölkerungsvermehrung,
verbunden mit einem progressivem Anstieg des sogenannten 'Lebensstandards' und
wachsendem Unverständnis für immer dringender werdende Naturschutznotwen-
digkeiten, führen zu multiplen Zivilisationsschäden, deren Indikatoren rücksichts-
lose Entwaldung, Immissionsbelastungen aller *Ökosysteme und Vernichtung
einzelner Tier- und Pflanzenarten sind. Einmalige tropische* Ökosysteme −
unterschiedlichen Alters −verschwinden, bevor sie wissenschaftlich erfaßt wurden,
für immer. Die tropischen Ökosysteme Brasiliens, gehören zu den artenreichsten
und in ihrer Einmaligkeit wertvollsten der Welt.

  Fast 90% Brasiliens liegt zwischen dem Äquator und dem südlichen Wendekreis.
Seine über 8.500 km lange Küste wird durch die südäquatoriale Variante des
Golfstromes, den mächtigen Brasilstrom, geheizt, so daß auch in den Wintermona-
ten im allgemeinen mittlere Küstenwassertemperaturen von über $18°C$ vorherr-
schen. Während die $20°$ Jahresisotherme durch Anden und kalten Humboldtstrom
im Westen S-Amerikas auf die Höhe des Amazonasstromes zu liegen kommt, wird
sie durch den Brasilstrom weit unterhalb des südlichen Wendekreises verlegt. Ihr

1 Die vorliegende Arbeit stellt den ersten Versuch einer ökologischen Gesamtgliederung
Ostparanás dar. Herr Prof. Dr. RALPH HERTEL, Chef des Botanischen Department (Curitiba)
stellte mir seine Kenntnisse beim Durchsprechen vegetationskundlicher Fragen uneinge-
schränkt zur Verfügung. Bei der Bearbeitung des geomorphologischen Teils konnte ich mit
stets freundlichem Entgegenkommen des Herrn Kollegen Prof. Dr. RIAD SALAMUNIS,
Direktor des geowissenschaftlichen Instituts unserer Universität rechnen. Beiden Herren bin ich
daher sehr zu Dank verplichtet, wie Herrn Prof. Dr. PAUL MÜLLER (Dept. Biogeography,
Universität des Saarlandes, 66 Saarbrücken, BRD) für die Diskussion biogeographisch-ökolo-
gischer Probleme.

Verlauf ist unter Mitwirkung anderer Hauptklimafaktoren für die Entwicklung und Abgrenzung komplexer Lebensgemeinschaften von entscheidender Bedeutung.

Der neotropische Riese Brasilien enspricht mit einer Gesamtfläche von 8.513.844 km² einem Lebensraume der ca. 34 mal größer als die Bundesrepublik Deutschland und selbst größer als der Australische Kontinent ist. Als tropenbiotopreichstes Land der Erde besitzt es dank seiner hauptsächlich aus Granit, Gneiss- und Glimmerschiefer bestehenden Urgebirgskerne (Kratone) ein festes Strukturfundament, welches seit Ende des Präkambriums die geomorphologische Entwicklung des Großraumes bestimmt. Zwischen diesen alten substratstabilen Festlandskernen ruhen die drei größten Tieflandsbecken S-Amerikas: Das Amazonas-, Sào Francisco- und Paraná-Uruguay-Becken.

Paraná hat als mittelgroßer Bundesstaat Brasiliens eine Fläche vond rund 201 201 km², ist also naheuzu 5 mal größer als die Schweiz und etwas kleiner als England. Seine Nord-Süd bzw. Ost-West-Ausdehnung (22°29'30'S − 26°42'59'S bzw. 48°02'24" − 54°37'38") wird von Flußebenen und Küstenstreifen begrenzt. Sehr übersichtlich ist auch der geologisch-geomorphologische Aufbau gegliedert. Die Serra do Mar teilt als Rumpf-Randgebirge das durch zwei Ingressionsbuchten zernagte tropische paranänser Küstentiefland von im Mittel um 900 m höher gelegenen zentralen, nach Westen geneigte Schichtstufenland ab, welches seinerseits nach Westen zum kontinentaltropischen Becken des Paraná-Stromes ausläuft. Dieses Schichtstufenhochland setzt sich geologisch und orographisch aus 3 sukzessiven Hochebenen zusammen:

1. Die altpaläozoische Hochebene Curitibas mit dem bergischen Land Açunguí und der Hochtafel Maracanà,

2. Die Hochebene Ponta Grossas mit ihrer ca. 35 km westlich Curitibas verlaufenden Devonstufe (= Serra do Purunà), welche in ihrem abfallenden Teil mesozoische Mesetenformation aufweist und

3. die Trapphochebene, mit ihrer Trias-Jura-Stufe (= Serra da Boa Espernaça), welche sich ihrerseits aus 4-5 Unterregionen zusammensetzt.

Die natürliche Landschaftsgliederung Paranás steht mit der ökologischen Raumdifferenzierung in direkter Beziehung. Es ist daher weiter nicht überraschend, daß auch die zoogeographischen Gliederungsversuche landschaftlich weitgehend mit den Großeinheiten Paranás übereinstimmen. So gehört der artenreiche tropische Küstenregenwald der zoogeographischen Provinz Tupí (CABRERA & YEPES 1940, FITTKAU 1969) und dem Serra-do-Mar-Ausbreitungszentrum im Sinne von MÜLLER (1973) an, deren natürliche Grenze nach Westen durch die Serra do Mar gebildet wird. Das artenarme paranänser Campo- und Grashochland von subtropischem Charakter gehört ausnahmslos zur Großprovinz Guarani (Parana-Ausbreitungszentrum). Angesichts dieser großräumigen Unterschiede geologischen Substrats, Klimaverhältnissen, Biozönosereichtums und räumlichen Ernährungsbedingungen eignet sich aber gerade der landschafts-ökologisch hochdifferenzierte Ostteil ganz besonders gut zur Entwicklung tropenökologischer Forschung.

Biogeographisch gehört Ostparana zur relativ isolierten Randtropenregion Brasiliens, genannt Tupí. Handelt es sich bei der Neotropis um eine biogeographische Region, welche die größte Zahl an eigenständigen Formen entwickelt hat, so können wir ganz besonders für das zwischen Mittelgebirge und Meer überragende Randtropensystem Ostparanas eine hohe Endemitendichte erwarten. Nach rezenten geomorphologischen Forschungen sind für diese Region wenigstens 2

erdgeschichtliche Trockenperioden zu unterscheiden, wo extrembiotopartige Ur-
waldinseln als urzeitliche Refugialbiotope das Überleben der ökotypischen
Feuchttropen-Biome gewährleistete. Zahlreiche geologische Strukturhomologien
zwischen unseren Festlandskernen (Serren) und ihren Afrikanischen Partnern
deuten auf enge Beziehungen in der Alt-Gondwanazeit hin. Andererseits gibt es
zahlreiche Hinweise für eine ehemalige Verbindung mit der heutigen Andenregion
mit Neuseeland. Während der letzten 10 Jahre sind zahlreiche Beispiele von
Grundwasserzönosen (syncariden) bekanntgeworden, welche auf eine heterogene
Bildung der südlichen Neotropis schließen lassen. Ich selbst konnte im Mai 1971
im süßen Grundwasser der Bucht v. Paranaguà eine *Parabathynellen*-Art nach-
weisen, deren nächste Verwandte nur noch aus dem Grundwasser Ostmadagaskars
sowie dem Osthang der Sierra de Cordoba bekannt sind. Ähnliche Verbreitungs-
bilder kennen wir von zahlreichen anderen Tiergruppen (i.a. Onychophoren,
Schlangen, Iguaniden). Angesichts des hauptsächlich erdgeschichtlich bedingten
N-S-Verlaufes der wichtigsten neotropischen Landschaftsgürtel verbunden mit
einem hohen Isolationsgrad ökologischer Einheiten, erwies sich die Erarbeitung
eines ökologischen Profils vom rein wissenschaftlichen wie auch unterrichtsprak-
tischen Gesichtspunkte aus, sehr geeignet. Gegenwärtig besteht auch aufgrund der
Tätigkeit von Entwicklungsexperten der FAO und UNESCO starke Nachfrage für
eine ökologische Kartierung Paranás.

Die Klimadiagramme für Curitiba (1. Hochebene) und Paranaguà (Küstentiefland) entsprechen Mittelwerten von mehr als 50 jähriger Temperatur
und Niederschlagsregistrierung der entsprechenden meteorologischen Stationen.
Folgende Charakteristika sind für die ökologischen Großräume bezeichnend: 1.
Die aus den Monatsmitteln gebildete Jahrestemperaturkurve liegt für das curitibaner Hochbecken nur ca. 4 Grad C tiefer als für das Tiefland. Während dies aber
total frostfrei ist, besteht für das Hochland um Curitiba während der Wintermonate (Mai — September) stets Frostgefahr. Frost aber bedeutet für viele Tiere
und Pflanzen den Tod. Allerdings kann das Thermometer an strengen Wintertagen,
auch im Feuchttropischen Tiefland, bis nahe an den Nullpunkt fallen und
umgekehrt sind frostfreie Winter, wie wir ihn dieses Jahr in Curitiba hatten,
durchaus nicht selten. Diese klimatischen Unterschiede wirken sich lokal jedoch
weit mehr aus als man durch die 4° Temperaturlinie erwarten könnte. 2. Die
Niederschlagsverhältnisse weisen jedoch mehr als die Temperaturverläufe bezeichnende Unterschiede für beide Großräume auf: Während das Hochland zwei unter
200 mm pluviometrische Maxima kennt (Oktober bzw. Januar) ist für Paranaguá
das Februarmaximum von mehr als 300 mm einzigartig. Das Tiefland kennt daher
keine Trockenperiode im Monat November, wie sie für das Hochland charakteristisch ist. Allerdings fällt auch im Tiefland die Niederschlagshöhe während der
Monate Juli und August beträchtlich unter 100 mm.

Die klimatischen Hauptunterschiede von Hoch- und Tieflandsraum sind
natürlich grundsätzlich durch die fast 2000 m erreichende Serra do Mar bedingt.
Von den Inseln der Bucht von Paranaguá aus gesehen, erhebt sie sich als eine
blaugrüne Wand, welche den Großteil der vom Meere herströmenden Feuchtluft
kondensiert und abregnen läßt. Nachtsüber hängen tiefe Nebelschwaden bis in die
Vorgebirgshügel hinein, welche während der Wintermonate erst gegen Mittag hin
von der Sonnenstrahlen aufgesogen werden können. Dieses Nebelwasser ist
ökologisch von großer Bedeutung, obgleich es pluviometrisch kaum erfaßt wird. Die
ökologische Hauptfunktion der Serra do Mar besteht also in der Trennung von

Hoch- und Tiefland. Sie wirkt als markante *Wasserscheide,* wobei sie verhindert, daß der Iguaçú-Fluß nicht direkt in den Atlantik mündet, sondern nach S und SW abgebogen wird, um dann an der Dreiländergrenze in mächtigen Fällen in den Rio Paranà zu münden. Da es sich bei der Serra do Mar um einen alten Festlandskern handelt, macht sich die trennende Wirkung hauptsächlich auch in der Besiedlung des Grundwassers beider Bereiche bemerkbar: Die interstitiellen Zönosen des Rio Nhundiquara, wohl der bedeutendste Tieflandsfluß des Paranänser Atlantik-beckens, haben völlig verschiedenen Charakter im Vergleich zu den entsprechenden Lebensgemeinschaften des Curitibaner Beckens. Der Grund- und Oberflächen-wasserscheide dieses Mittelgebirges geht aber auch die trennende Wirkung hinsichtlich Bodenbeschaffenheit und Vegetationsdecke parallel. Die ökologische Tiefenwirkung ist erdgeschichtlich, edaphisch, klimatisch, faunistisch und flo-ristisch weit größer als man auf den ersten Blick hin erwarten könnte. Sie führte zur Bildung zahlreicher tropischer Lebensräume in rascher, gesetzmäßger Folge. So schneidet der 25. Breitengrad zwischen dem Curitibanen Becken und dem Küstenwasser vor Paranaguá nicht weniger als 17 gut definierbare natürliche Einheiten, die wir im Folgenden als Makro-Ökosysteme beschreiben möchten.

1. Der Kontinentalschelf

Die interstitiellen Sandzönosen sind faunistisch außerordentlich konstant. Sie stellen eine, obgleich aus sehr verschiedenen Tiergruppen zusammengesetzte, einheitliche Lebensgemeinschaft dar. Sie besiedeln die gerade vor der Küste Paranás weit ins Meer hinausreichende Kontinentalschelfplatte. Dieses als subma-rine Ausdehnung des Küstenstreifens von uns seit Jahren untersuchte Ökosystem erreicht seine maximale Ausdehnung von mehr als 200 Seemeilen gerade vor Paraná. Dank seiner hauptsächlich aus Lehm-, Sand- und Kiesformationen von bestimmtem granulometrischem Werte (1-4 mm φ) bestehenden Biotopen, ist dieses Ökosystem auch geophysikalisch als ungewöhnlich resistent und dauerhaft anzusehen, was seine Bedeutung als solides Nährsubstrat für die Epifauna nur bekräftigt.

Dieses interstitielle Nährsubstrat besteht hauptsächlich aus Detritisfressern aller Art, wie hochkonzentrierte Protozoenpopulationen, hauptsächlich Ciliaten, aber auch multiple Copepodenassoziationen harpacticoiden Typus zusammen mit zahlreichen Nematoden-Gastrotrichen- und Micropolychaeten-Arten. Trotz seiner grundlegenden ökologischen Bedeutung wurde seitens der Ozeanographen den Kontinentalschelflebensgemeinschaften kaum nennenswerte Beachtung geschenkt. Sie stellen aber im Gesamtlebenszyklus der Küstenmeere ein wichtiges Nahrungs-bindeglied innerhalb der Stoffwechselkette von minderwertigen zu proteinreichen hochwertigen Nahrungsträgern dar. Es ist daher nicht verwunderlich, daß die zwischen Kap Frio und Kap Santa Marta gelegenen Kontinentalschelfgründe beachtliche Fischereierfolge aufweisen. Mehr als 50 Fischarten, die hier leben, sind von wirtschaftlicher Bedeutung. Ihre Phänologie wurde in den Jahren 1964-1968 von uns eingehend untersucht. Auch der Krebsfang ist beachtlich und kann noch sehr ausgebaut werden: Penaeiden, Garneelen, Langusten und Krabben aber auch Schildkröten leben direkt auf den sandigen Schichten und zwischen den Steinen des Kontinentalschelfs, immer auf Lauer, nach der interstitiellen Kleintierwelt.

Abgesehen von Biokonstanz und Geokonstanz zeichnet sich dieses Ökosystem noch durch eine ungewöhnliche Thermokonstanz sowie Chemokonstanz aus. Während in unseren Breiten das Küstenmeeresoberflächenwasser während des Jahres zwischen ca. 18 und 28 Grad variieren, sind die Temperaturschwankungen auf der Oberfläche des Schelfgrundes wesentlich geringer (21-25°C). Dies bedeutet aber, daß die Meiofauna sich hier unter Treibhaustemperaturen das ganze Jahr über fortpflanzen und ausbreiten können. Eine winterliche Unterbrechung der Lebenszyklen gibt es in den Schelfböden niedriger Breiten, die noch dazu durch Warmwasser (Brasilstrom) ständig geheizt werden, nicht.

Unwesentliche Schwankungen gelten auch für Salzgehalt und pH des Schelfwassers. In Tiefen von mehr als 10 m wirken Sandbänke als eine Art Schwamm, welche relativ große Mengen Salzwasser praktisch unabhängig von den Gezeitenströmungen zurückhalten kann. Lediglich die Oberflächenschichten werden durch Strömungen oder Nekton aufgefüllt und gewaschen. Meereswasser mit höheren Salzgehaltswerten und Dichtewerten wird auch in der Nähe von Buchten und Flußmündungen (B. de Guaratuba und Paranaguá) zur Niedrigwasserzeit zurückgehalten und widersteht sogar dem gelegentlich durch starke Regen verursachten Salinitätssturz des Oberflächenwassers, welches anläßlich der plötzlichen osmotischen Dystonie zu Massensterben von Zoo- und Phytoplankton-Zönosen führen kann.

So entwickeln sich die Zönosen des Kontinentalschelfs unter äußerst günstigen Umweltbedingungen, welche hauptsächlich auf den Einklang von physikalisch chemischen Parametern zurückzuführen sind, eine Tatsache, welche gerade für die Küstengewässer unserer Breiten als charakteristisch angesehen werden kann. Die Zusammensetzung aus Sanden verdankt der paranänse Schelf in erster Linie dem geologischen Aufbau der Küstenformation. Die sog. Alexanderformation des Pleistozäns, welche zum Aufbau sedimentären Ursprungs des Kontinentalschelfs wesentlich beigetragen hat, zusammen mit den alluvialen Ablagerungen, garantiert ausgedehnte Sandformationen zur Erhaltung submariner Ökosysteme des Sublitorals.

Die Bedeutung des südbrasilianischen Schelfs für die einheimische Fischerei wurde nicht nur durch Arbeiten der beiden brasilianischen Ozeanographischen Forschungsschiffe, sondern auch von nordamerikanischen, japanischen und jüngst auch durch das deutsche Fischereiforschungsschiff WALTER HERTWIG bestätigt. Die am 2. Juni 1971 in Kraft getretene Erweiterung des brasilianischen Hoheitsgebietes auf 200 Seemeilen stellt also die Meeresfauna des Kontinentalschelfs sozusagen unter offiziellen Schutz, nachdem sie gerade in den letzten Jahren einer erschreckenden Raubfangtätigkeit ausgesetzt war. In den Tiefen des Kontinentalschelfs sind auch hier in Paraná ansehnliche Erdölvorkommen zu erwarten. Spezialbohrtürme der PETROBRAS werden in nächster Zeit hierüber Klarheit schaffen.

2. Das Thalassopsammal (Strand- und Küstengrundwasser)

Die Erforschung dieses Ökosystems gehört zu einem der interessantesten und modernsten Kapitel der Meeresbiologie. Es handelt sich wie überall in der Welt auch hier um typische Übergangsbiotope, deren Hauptsubstrat feuchte Strandsandschichten sind. Der Salzgehalt fällt hier innerhalb weniger Meter, nach Überwinden des Tidenbereiches, auf Süßwasserkonzentration ab. Mit ihr verändern sich entsprechend der pH, die biozönotischen Zusammensetzungen der Interstitialfauna sowie der speziellen Ernährungs- und Grundwassertemperaturverhältnisse. Neben der Breitstrandküste, welche die Bildung echter Küstengrundwasserräume erst ermöglicht, kommen in geringem Maße auch Kliffbildungen und Felsenformationen die direkt ins Meer tauchenden Küstenberge vor (Caiobá, Guaratuba etc.). Charakteristisch ist allerdings die Breitstrandküste mit anschließender mehr oder weniger breiter Dünenzone. Hier im strandigen Kontinentalrandgrundwasser finden sich auf nur ca. 30 m Breite euhaline, polyhaline, mesohaline und olygohaline Unschärfebereiche, wie wir sie analog im Oberflächenwasser der Buchten für eine Ausdehnung von mehr als 30 km kennen. Dieses Küstengrundwasser mit äußerst hohem Salzgehaltsgradienten steht ständig, aber ganz besonders während der Regenmonate Januar und Februar, unter hohem hydrostatischen Süßwasserdruck, dank der relativen Nähe der Serra do Mar, welche für eine stete Überhöhung des phreatischen Niveaus mit überreichem Grundwasserangebot sorgt. Hierauf ist wohl teilweise zurückzuführen, daß schon in wenigen Metern von der Flutlinie weg Trinkwasserbrunnen, die kaum verbrackwassern, angelegt werden können. Diese weisen dann eine rein limnische Grundwasserfauna auf, während Grabungen im nur wenige Meter auswärts gelegenen Tidenbereiche zum Auffinden arten- und individuenreicher Thalassopsammalzönosen führen können. Wo Granit- Gneissmasive als Küstenberge ('Zuckerhüte') ihre Felswände direkt ins Meer tauchen, finden wir Assoziationen von Meeresalgen (Chlorophyceen, Phaeophyceen und Rhodophyceen) und Kleintieren (Fische, Crustaceen, Nematoden, Polychaeten, Seepocken usw.) welche an Stellen, wie z. B. in Cabo Frio, wo sie den Gesamtküstencharakter bestimmen, als eigenes Ökosystem, das litorale Felsbenthal oder modern als Thalassopsephal behandelt werden können. Hier in Paraná haben diese als tidalsublitoral zu bezeichnende Biozönosen, welche submarin die untertauchenden Felsen bedecken, nur sehr untergeordnete Bedeutung. Nur in Ausnahmefällen, wie z.B. der Ilha das Palmas sind sie in ihrer ursprünglichen Form vereint mit Miesmuschel- und Austernbänken noch anzutreffen. In den Vorbergen, wie denen von Matinhos, Caioba, Guaratuba, Brejatuba und Saí, sind sie angesichts der progressiven Verschmutzung praktisch verschwunden. Lediglich Ulvaceen mit kümmerlichem Austern- und Miesmuschelbelag zusammen mit sessilen Polychaeten erinnern noch an ausgedehnte ursprüngliche artenreiche Bioregionen. Stets verschwinden diese Formationen mit dem Felsensubstrat alsbald unter der Kontinentalschelfdecke, so daß sie nicht die ökotypische Voraussetzung zur Bildung eines eigentlichen substratgebundenen Ökosystems hier in Paraná darstellen.

Im Gezeitengürtel finden wir selbstverständlich reichliche, hauptsächlich aus Mollusken, Kleinkrebsen und Polychaeten zusammengesetzte Zönosen des Endopsammals und Epipsammals. Manchmal kommt es zur Dominanz relativ seltener Endopsammalvertreter, wie z. B. des wurmförmigen Hemichordaten *Balanoglossus* sp. – oder des sandbohrenden Dekapodenkrebses *Emerita emerita* (Tatuí), welche allerdings angesichts der offenbar irreversiblen Verschmutzung unserer Strände zur ausgesprochenen Seltenheit verdammt sind.

3. Restingal und Dünenzone

Obgleich Restinga und Dünenzonen zwei gut trennbare Ökotope sind, werden sie angesichts der charakteristischen Verknüpfung an der paranänser Küste gemeinsam behandelt. Auch fehlt vielerorts die eigentliche Dünenzone. Andererseits können Bergdünen, wie sie im Süden Paranás und vor allem auf der Insel Santa Catarina (Lagoa-Conceição) vorkommen, die Restinga übersanden. Restingal kann man als Strandgehölz übersetzen, doch möchte ich die Gelegenheit nicht versäumen, auf den weit höheren nomenklatorischen Wert der Bezeichnung Restingal hinzuweisen. Er sollte forthin von den Ökologen bewußt angewendet werden. Ich versuche daher hier eine einschlägige Definition zu geben: Unter Restingal versteht man den tropischen, windflüchtigen Strandgehölzgürtel des Supralitorals in hügeliger Landschaft, welcher vom Breitsandstrand mit Dünenwall zum Küstenhochwald überleitet. Zwischen der eigentlichen Restinga und den charakteristischen Tropen-dünenformationen bestehen auch in Paraná viele Übergänge. Anwandlungsstufen, von der niedrigen, substratgebundenen Form einschichtiger Strandlage bis zur mehrsichtigen komplexen Restinga mit Epiphytenbesatz können auftreten. Die auf salzarmen Sandböden wachsenden Formationen zeigen die charakteristische Filz-wurzelschicht in welche dicke Krüppelwurzeln eingelassen sind. Strukturell lassen sich meist 3 Lagen unterscheiden:

1. Die Strauchschicht mit folgenden ökotypischen Pflanzen: *Psidium littorale, Psidium guajava*, das Edelholz Jacarandá, *Schinus terebenthifolius, Erythroxylum amphifolium* (ein guter Windbrecher), *Noranthea brasiliensis, Cedrela* sp., Ilex-Arten. Im kargen Boden findet doch eine Art Autohumifizierung statt und Bildung einer charakteristischen Wurzelschicht.

2. Die Krautschicht: Als typischen Humusmacher *Polystichium adiantiforme, Blechnum serrulatum, Polypodium latipes*, und sogar Orchideen, *Epidendrum latilcabre* sowie *Cleistia* sp.

3. Epiphytenschicht: Häufig ist hier *Polypodium vaccinifolium, P. percussum, Pleurothallis pedencularis, Aechmea nudicaulis, Vriesia carinata, Tillandsia tenui-folia* sowie zahlreiche Lichenes und Bryophyten, welche noch einer botanischen Bearbeitung harren.

Die faunistische Zusammensetzung dieses Ökosystems ist ebenfalls noch nicht untersucht. Doch findet man ohne Schwierigkeiten im Substrat psammophile Käfer eurylope Schaben, Dermapteren und Maulwurfsgrillen sowie an den feuchteren Stellen Pauropoden, Symphilen, Asseln und Colembolen, um nur einige Tiergruppen zu nennen. Unter der Macrofauna erweisen sich standortspezifisch Schlangen, Erdleguane (*Liolaemus occipitalis*) und im Holzinterstitium nistende Kolibris ökologisch als besonders interessant. Multiple Wechselbeziehungen zwischen Milben, Termiten, Ameisen und Aphiden harren noch einer wissenschaft-lichen Untersuchung, von den zahlreichen Tausendfüßlern, Pseudoskorpionen und Spinnen ganz abzusehen.

4. Der Küstenhochwald (Epiphytal u. Phytothelmal)

Dieses Makroökosystem zieht sich entlang der Küste durch ganz Paraná. Wo felsige Berge den feuchten Küstenwald ansteigen lassen, verhält er sich als dichter Streifen, um sich dann auf den Sandniederungen, z.B. zwischen Praia de Leste und Pontal do Sul oder auf den Flachinseln des Buchtinnern (Ilha do Mel) nach innen hin aufzulockern. Je nach den edaphischen Bedingungen können unterholzreiche bzw. unterholzarme Küstenhochwaldformationen unterschieden werden. Auf relativ trockenen Grunde dominieren *Cabrulea* sp. *Hieronima, Nectandra, Schizo-lobium, Pindaria, Pachyra* während *Erythrina, Hedychium, Marliera, Calophyllum* und *Bactris* mehr für Feucht- bzw. Sumpfbiotope charakteristisch sind. Obgleich die Zersetzungsvorgänge im feuchttropischen Innern dieses Ökosystems besonders während der heißen Sommermonate quantitativ und qualitativ sehr hoch sind, kommt es nur in besonders günstigen Lagen wegen der meist hohen Umwelt-temperatur zur Humusbildung.

Dem gegenüber steht aber die geradezu verblüffend üppige Pflanzenwelt, trotz praktisch kontinuierlicher Auswaschung des Bodens. Dies wird gerade für den Küstenwald zwischen Pontal do Sul Praia de Leste sehr augenscheinlich. Dieser Streifen zeichnet sich auch noch durch ein reichliches Wassernetz aus. Nach Ursprung, Wasserführung und Wasserqualität können wir wenigstens 2 Gruppen unterscheiden. Flüsse, die autochthon im Küstenwald selbst entspringen (= 'gamboas') und solche, welche aus dem Vorlande der Serra do Mar kommen ('nascentes'). In Paraná müssen aber die ursprünglich sauerstoff- und mineralsalz-reichen Nascentes erst das Pantanal (Ökosystem Nr. 10) durchlaufen. Daher kommen die anfänglichen Unterschiede mündungswärts praktisch zum Verschwin-den, was sich auch aus der gleichförmigen Urwalduferformation der Küstenhoch-waldwässer ergibt. Dem Hydrobiologen interessieren ganz abgesehen von den schwarz- bis grüngelben Tieflandsflüssen hauptsächlich die Biotope der Nischen-wässer des Epiphythals (vorwiegend Trichterwasser v. Bromeliaceen und Araceen), sowie der Blattachseln im besonderen (= Phytothelmal), welche gerade im feuchten Teil dieses Ökosystems reichlich vertreten sind. Zahlreiche aus diesen pflanzlichen Interstatialwässern untersuchten Proben erwiesen, daß ihre Microfau-napopulationen hauptsächlich aus Copepoden, Ostrakoden, Nematoden, Oli-gochaeten aber auch aus Chironomiden und Ephemeridenlarven zusammengesetzt sind. Das meeresnahe Epiphytal zeigt innerhalb jener Gruppen außerdem Übergänge zwischen halophilen Trichterpopulationen und reinen Süßwasserbio-zönosen. Eine Gesamtuntersuchung dieser so interessanten Phythalbewohner steht allerdings noch aus. Vielleicht sind diese Trichterwässer wenigstens zum Teil dafür verantwortlich, daß trinkwasserlose kleine Meeresinseln ohne Schwierigkeiten von einer Vogelfauna und Amphibien besiedelt sind, welche sich dieses gesammelten Regenwassers als Trinkwasserreservoir bedienen können. Kaum trifft man echtes Epiphytalwasser stinkend oder verfault an. Die multizelluläre Kleintierwelt spielt als Bindeglied zwischen den zahlreichen Protozoen und Bakterien eine wichtige Rolle bei der Autopurifikation jener Trichtersubstrate, die im Trichtergrund oft sogar Sand enthalten, welche höchstwahrscheinlich durch Vögel, Baumfrösche und Kröten[1] eingeschleppt werden dürften. Allerdings hat dieses Epiphytal seine eigene Insektenfauna und die zahlreichen Baumfrösche laichen gerne in jenen Trichtern, wo ihre Kaulquappen dann die Verwandung durchmachen.

5. Das Tholomigal (Buchtweichböden)

Es besteht meines Wissens noch kein Eigenname in der Terminologie dieses Öko-systems, eines der bestabgrenzbaren, die es gibt. Ich schlage daher gleichzeitig vor, in der Ökologie den international und wissenschaftlich klar erfaßbaren Ausdruck

---

1 Landplanarien/Oligochaeten.

Tholomigal zu gebrauchen. Der Bodenschlamm reich an Seston der Buchten setzt sich aus terrigenem aber auch thalassogenem Material zusammen, welche im Buchtbecken durch die Gezeitentätigkeit einer ständigen Waschung, Sonderung und Verlagerung unterworfen sind. Das tropische Tholomigal weist eine Reihe Sonderheiten auf, die insbesondere durch die relative Temperaturkonstanz mit Fortpflanzung der Populationen zu allen Jahreszeiten sowie das überreichliche Angebot an organischer Substanz aller Art bedingt sind.

Da sich die brasilianische Küste seit Ende des Tertiärs in steter Hebung befindet, zeigt der Küstenstreifen auch weiterhin wachsende Tendenz. Verlandungsprozesse sind überall anzutreffen und stets wandeln sich festlandsnahe Tholomigalstreifen in fruchtbare Erde um.

Trotz relativer Temperaturgleichheit ist die Fauna dieses Ökosystems als arm zu bezeichnen. Es kommen hauptsächlich Würmer aller Art vor (Tubificiden, z.B. *Chaetopterus variopedatus*), hauptsächlich wenn Sandformationen am Aufbau des Substrats teilnehmen. An günstigen Stellen lassen sich allerdings gute faunitische Strukturen feststellen, welche von zahlreichen Protozoenarten bis zu Fischen, über Crustaceen, Insekten und Mollusken, wie z.B. dem Buchtschollen reichen *(Paralichthys brasiliensis)*. Copepoden und Nematoden sind im Tholomigal durch zahlreiche typische Arten vertreten, welche verständlicherweise euryhalin sind. Ökologisch sehr interessant sind die während der Ebbe freiliegenden Tholomigalstreifen, welche von Wasservögeln aller Art besucht werden, da sich dort außer der eigenständigen Fauna auch noch viel durch Flutwasser angeschwemmtes Getier aufpicken läßt, welches oft schon angeschlagen, durch das schwammige Schlammsystem zurückgehalten wurde.

So gibt es ganze aus Tholomigalmaterial zusammengesetzte Buchtinseln, welche bei Ebbe aufkommen, dann durch das Flutwasser für einige Stunden wieder bedeckt werden, um von neuem Meeresmaterial anzuhäufen.

Die vorliegende ökologische Einheit ist jedoch keineswegs mit der Mangroveregion zu verwechseln, obgleich selbstverständlich auch hier eine Übergangszone besteht. Beiden ist gemeinsam, daß sie zum großen Teil zum oligo-polyhalinem Flachküstenwassergebiet gehören, wo unter Brechung der osmotischen Resistenz ein ständiges rhythmisches Sterben des Planktons mit nachfolgender Sedimentation vor sich geht.

6. Das Inselmangrovial

Dieses Inselsystem ist wissenschaftlich nur noch von den Rosario-Inseln Kolumbiens bekannt, also aus der Karibischen See, wo *Rhizophora sp.* Korallenbänke besiedeln. Auch im SW Puerto Ricos bei La Parguera und Mayaguez sind ähnliche Formationen mit Korallenbänken assoziiert zu finden. Sämtliche Inseln der Bucht von Paranaguá weisen Mangrovegürtel auf.

Diese sind meist im Südostteil der Inseln entwickelt. Allerdings sind hier keine Korallensubstrate vorhanden, welche sich in Brasilien nur bis zum 15. Breitengrad S entwickeln. Auch sinds Biomasseverhältnisse, Salzgehaltsschwankungen und Windverhältnisse sehr von denen der Festlandsrandmangrovialen unterschieden. Sie sind tiefgreifenderen Extrembedingungen, wie plötzliche Flutkatastrophen ausgezetst als die letzteren. Substratvarianten, wie plötzlicher Felsengrund oder Sandstreifen wechseln mit rein schlammigen Grund im Inselgürtel ab. Selbst das

Mangroverandgras *Spartina brasiliensis* weist eine Inselvarietät auf. Seine Filzlagen tragen wesentlich zum Zurückhalten von Flutmaterial bei. Die Rasenbildung scheint bei den Insel-Spartinaformationen dichter zu sein, als bei den die Kontinentalränder säumenden Meerwassergräsern.

7. Der Inselwald

Seine Bildung hängt in erster Linie von Größe und Grund der jeweiligen Insel ab. So weisen flache, aus alluvialem Material gebildete Großinseln, wie z.B. die Ilha do Mel, praktisch nur Restinga-Formationen auf. Andere Inseln der Bucht von Paranaguá, deren Sedimentsgürtel sich um ein ca. 150 m hohes Granit-Gneissmassiv (Beispiel: Ilha da Cootinga) gruppieren, konnten dank ihrer Kratonzugehörigkeit bis heute, aller Menschenhand zum Trotz, einen charakteristischen, geschlossenen Inselwald bewahren. Ihre floristische und auch teilweise faunistischen Biozönosen sind praktisch noch nicht untersucht, und es besteht dringende Gefahr, daß sie in wenigen Jahren verschwinden, bevor sie wissenschaftlich erfaßt werden konnten. Diesen alten Inselkopfwäldern stehen die kaum als Wald anzusprechenden Baumvorkommen von erst kürzlich aus dem Meere aufgetauchten Inseln gegenüber.

8. Inselpsammal- und Inselpsephal

Fast jede größere, massive Buchtinsel weist hier auch eine Sand- bzw. Schotterzone auf, welche auf der Insel do Mel, der Insel da Cootinga sowie der von uns eingehender untersuchten Insel do Teixeira auf der Nord- bzw. Nordostseite der Insel liegen. Im Gegensatz zum bereits oben beschriebenen typischen Thalassopsammal des Meeresstrandes handelt es sich hier praktisch um ein, obgleich von Brackwasser umflossenes, ausgesprochenes Limnopsammal. Seine Korngröße kann sehr variieren. Wo ansehnliches Limnopsammal auf Inseln vorkommt, kann stets mit reichlicher Grundwasserführung gerechnet werden.

Den Siedlern fehlt es daher nicht an Trinkwasser und so ist es weiter nicht erstaunlich, daß gerade diese Streifen der Inseln am reichsten und seit den ersten Anfängen der portugiesischen Kolonisation besiedelt sind. Zahlreiche auf den Inseln Cootinga, Mel und Teixeira durchgeführte Wasseranalysen ergaben, daß es sich um rein süßes Grundwasser handelt mit Salzgehalten unter 0.5 mg%. Dank der hohen Niederschlagstätigkeit enthält das interstitielle Sandsystem jener Inseln ein ausgesprochenes phreatisches Niveau mit echtem nie versiegenden Wasserreserven. Dieses läuft dann in das aus Kiesel- und Glimmerschieferschollen bestehende sog. Psephal aus, welches ganz besonders eindrucksvoll am Nordrand der Ilha do Teixeira, ganz im Innern der Bucht von Paranaguá zu sehen ist. Hier wird dem Kenner auch richtig bewußt, daß es sich wirklich um ein eigenes Ökosystem handelt, welches sich gerade durch den ausgesprochenen Reichtum an Nischen aller Art, welche von Krebsen, Fischen, Mollusken besiedelt sind, auszeichnet.

Aber gerade die faunistische Zusammensetzung des grundwasserführenden Inselpsammals läßt auf alte erdgeschichtliche Zusammenhänge mit dem Hauptkraton der Serra do Mar schließen: ihre biozönotische Lebensform ist als typisch eucaval zu bezeichnen, da wider Erwarten, echt limnische Grundwasserkrebse vorkommen, welche sonst nur in archaischen stygobionten Lebensgemeinschaften des Festlandes angetroffen werden: Lebende Fossilien der Ordnung der Syncariden, welche in Extrembiotopen vereinzelt auf den 5 Kontinenten anzutreffen sind, sind Zeugen einer mehr als 300.000.000 Jahre alten Süßwasserfauna. Den hier auf der Ilha do Teixeira und Ilha da Cootinga gefundenen Vertreter nannten wir *Parabathynella paranaensis* n.sp. Seine nächsten Verwandten kommen nur noch am Osthange der Sierra de Cordoba (Argentinien) sowie im Nordosten der Großinsel Madagaskar vor. Wahrscheinlich ist das Vorkommen dieser 3 Arten direkt an die Erhaltung der entsprechenden drei Altgesteinsblöcke der Gondwanaperiode gebonden, welche seit dem Altpalaeozoikum allen Gewalten zum Trotz sich erhalten konnten. Die sie begleitenden Psammal- und Psephalformationen konnten daher als physikalisch und chemisch relativ konstante Substrate eine der altertümlichsten Grundwasserzönosen bis in unsere Zeit herrüberretten. Sehr interessant ist, daß diese archaischen Zönosen, trotz intensiver Suche, nur auf den beiden größten Altinseln der Bucht von Paranaguá gefunden wurden, welche durch poröse Sand- und Sandsteinschichten des Pleistozäns (sog. Alexanderformation) mit den homologen Limnopsammalbiotopen des Festlandes dauernd in Verbindung stehen. Im Hochland der Serra do Mar kommt ja algonkischer, höhlenreicher, karstähnlicher Kalk- und Dolomit-Gürtel vor (Serra 'Açunguí'), welche in phreatischem Niveau seit uralten Zeiten reichliche Grundwasseransammlungen ermöglichten. Diese üben seither einen beachtlichen hydrostatischen Druck auf die interstitiellen Ökosysteme des Küstenvorlandes und seiner Inselwelt aus, was im Einklang mit dem druck- und spaltenreichen Unterbau die Permanenz alteingesessener Süßwasserrelikte nur begünstigte. Gleiche Interpretationsprinzipien gelten für die unter dem Namen Mesopsephal bekannten Grobschotterlagen, deren Bildung durch den Glimmerschiefer begünstigt ist.

Das Vorkommen der sog. Piçarra-Formation, wie wir sie vom Norden der Ilha do Mel her kennen, verhindert eine ständige Verbindung mit den eigentlichen unterirdischen Gewässern.

Der mitunter steinharte Picarrit ist hauptsächlich aus Eisen und Aluminiumhydroxyd zusammengesetzter Latosol, also undurchlässig. Die darauf ruhenden Bleichsande werden ständig von humussäurehaltigem Regenwasser, welches die Rhizo-Filzschicht durchsickert, gewaschen. Da aber das sich ansammelnde Sickerwasser nicht die Picarritschicht durchdringen kann, zeigen solche Böden Tendenz zur Moorbildung, was auch allenthalben in der Bucht von Paranaguá festzustellen ist.

Tab. 1. Physikalisch-chemische Daten des Wassers einiger Inseln der Bucht von Paranaguá.

| Insel | Salzgehalt S‰ | | Temperatur pH | | | |
	Bucht	Brunnen	Bucht	Brunnen	Bucht	Brunnen
Cootinga	30,5	0,2	19	22,5	8,5	6,0
das Cobras	24,7	0,3	20	22,0	8,5	4,5
Gerere II	19,8	0,1	22	24,0	8,0	7,0

9. Das Mangrovial

Nach Substrateigenheit, Nahrungsgrundlage, Klimasonderheiten und Lebensgemeinschaften handelt es sich um eines der gut charakterisierbaren tropischen Küstenökosysteme überhaupt. Ausschlaggebend für ihre Verbreitung ist die mittlere Jahrestemperatur. Die südliche Grenze der Mangroveverbreitung wird in S-Amerika von der $20°$ Jahresmittelisotherme erheblich beeinflußt. Auch andere rein physikalische Voraussetzungen, wie z.B. das völlige Fehlen einer Brandung, müssen bestehen, damit sich ein Mangrovial bilden kann.

Da die brasilianische Küste sich in geologischer Hebungsphase befindet, entstehen immer wieder neue Mangrovewälder, die alten werden meist vertieft und erweitert. Allerdings hält auch hier der Mensch Raubbau und holzt ganze Gebiete ab. In der Bucht von Paranaguá stellen die Altmangrovegebiete willkommene Fischplätze dar. Zahlreiche Fischer versuchen immer wieder wegen des Fischreichtums und relativen Geborgenheit gegen Stürme sich im Mangrovial seßhaft zu machen, was aber schon allein wegen des Mangels an Trinkwasser sowie der lästigen Insektenplage zum Scheitern verurteilt ist. Sie siedeln bald auf echtes Grundwasser führende Festinseln über, wovon aus sie dann ihr Mangrovejagdgebiet aus betreuen.

Im Gegensatz zum Inselmangrovial, welches zahlreiche endemische Organisationseigenheiten aufweist, bildet das Festlandsmangrovial weite Buschformationen, welche flußaufwärts und auch buchteinwärts sich zu Hochwald entwickeln können. Im Flutbereich der Ästuarien wachsen die meist ovalen Mangroveinseln. Nach Beseitigung des Mangroveinselholzes wird der durch die Wurzelschicht gehaltene Schlickboden alsbald Opfer der oft reißenden Gezeitenströmungen (zahlreiche Beispiele zwischen Parnaguá und Antonia). Hauptart ist die stelzwurzelige, vivipare und pneumatophorentragende *Rhizophora mangle*, welche hier im Zuge der allgemeinen Verlandungsprozesse durch *Laguncularia racemosa* und *Avicennia tomentosa* ersetzt wird. Andererseits kommen *Conocarpus erectus* und *Pelliciera rhizophora*, welche ich im Süden Puerto Ricos antraf, hier nicht vor.

Physikalisch-chemisch zeichnet sich das Mangrovial durch extreme Schwankungsbreiten aus. Während der Wintermonate sind nicht nur die Tages-Nachtschwankungen der Lufttemperatur sehr beachtlich (ca. $10-25°$ im Schatten), sondern selbst die Gezeitenströmungen tragen auch zu erheblichen Temperaturschwankungen bei, wie man im lagunären System Paranaguás leicht registrieren kann. Während eines sommerlichen Winternachmittags bis auf $28°$ erhitztes Mangrovepfützenwasser kann während der Nachtflut ohne weiteres bis auf $18°C$ heruntergekühlt werden, um nach einem darauffolgenden Sturzregen aus der Serra do Mar, sogar auf $10°C$ zu fallen. Schwankungen um mehr als 100% gehören durchaus nicht zu Seltenheiten. Ganz ähnlich verhält sich auch der Salzgehalt, da das Mangrovial ja hydrobiologisch zum Oligohalinal, Mesohalinal bzw. Plyhalinal gehört. Teilweise mag xerophytische Blattbildung, allgemeine Armut des Unterholzes und der Artenzahl der Bodentierwelt, hierdurch zu erklären sein. Mangrove dringt in Flußmündung aufwärts vor, so z.B. bis dicht an die Stadt Morretes, welche ca. 50 km. vom offenen Meere, also weiter von diesem als von Curitiba, entfernt liegt. Ihre Verbreitung wird durch die Flutwelle begrenzt. Landeinwärts dominiert im Unterholz der mannshohe Farn *Acrostichum aureum* von rosettenförmigem Wedelwuchs.

10. Das Pantanal

Die kontinuierlich seit dem Tertiär andauernde tektonische Hebung der Küste Südbrasiliens, führt allenthalben zu Flußverlegungen, Lagunen- und Sumpfbildungen. Die über 300 Jahre alte, oben bereits erwähnte Stadt Morretes, welche noch vor ca. 20 Jahren zu den gefürchtesten Malariaplätzen S-Amerikas gehörte, liegt nur wenige Meter über dem Meeresspiegel (Bahnstation mit 9 m ü.M.), auf einer Sandplatte des sumpfigen Ökosystems. Fast alljährlicht tritt der Hauptküstenfluß Nhundiaquara über die Ufer, wobei Regenüberschuß im Sommer, verbunden mit Sturzfluten, Hand in Hand gehen.

Immerfeuchte Zuckerrohrplantagen wechseln mit *Hidychium-Typha* Sümpfen einerseits, andererseits aber mit typischem noch bestehendem Sumpfurwald ab. Unter Wasser stehende Reisfelder sind lediglich im Nachbarstaat Santa Catarina häufig und beherrschen das Landschaftsbild der sumpfigen Tiefebene. Der Sumpfwald charakterisiert sich durch eine oft mannshohe Bryophytenschicht, das Vorkommen vieler Hochgräser, *Blechnum*-Arten, Sphagum-Polystichum und Funaria-Assoziation bilden Tiefmoorformationen. Im eigentlichen Sumpfwald finden wir *Erythrina* sp., *Tabebuia* sp., welche mit Bromeliaceen, Araceen und Orchideen überhangen sind. In den stehenden Sumpfgewässern sind als Schwimmpflanzen *Eichhornia* sp., *Salvinia* sp., *Azolla, Wolffia* und Lemnaceen häufig. Die auf den leichten Erhebungen stehenden Strauchformationen schließen zahlreiche Hochgräsern ein (*Mayaca sellowiana*). Auch findet man allenthalben die Kleinstechpalmen *Bactris* sp. und *Astrocarium*, oft unterwachsen von *Hidychium* sp. und *Typha* sp. Bänder von Hochgräsern, Cyperaceen und Liliaceen säumen den Bruch ein. Dieser Pantanalwald wird von zahlreichen Insekten bewohnt, die zum großen Teile noch unbeschrieben sind. Sie sind Amphibien (*Bufo ictericum*), Großeidechsen (*Tupinambis teguixin*) und immer noch gelegentlich anzutreffenden Jacares (*Caiman latirostris*) eine willkommene Nahrung. Eine typische Vogelwelt pflanzt sich hier fort, wie z.B. die Anhuma-Ente (= Urwaldwache), welche 2 knöcherne Sporen an jedem Flügel hat, neben der Capororoca Schwan.

Im Bruch sind Steißhühner (Crypturi), Saracuras und Hokkohühner nicht selten, um nur auf einige Vertreter aufmerksam zu machen.

Der Boden ist oft relativ reich an Humussäure auf sandigschlammigem Grund, mit nur trägem Sickerwasser. Echtes Grundwasser ist nur auf den Sandlinsen vorhanden, was siedlungsgeographisch bedeutungsvoll ist. Im Bruch und Ried müssen sich die Siedler mit detritusreichem, unhygienischem Oberflächenwasser abfinden, was zu zahlreichen Darm- und Lebererkrankungen führt. Das Gleiche gilt mehr oder weniger für die Haustiere. Nur allmählich macht sich ein hygienisches Bewußtsein auch unter der Landbevölkerung breit.

11. Bananal (= Tropenkulturebene)

Im Bananal, ist die Banane zweifellos die Hauptkulturpflanze einheimischen Ursprung. Landschaftsprägend neben ihr stellen Zuckerrohr, Mais, Manihot und viele Tropenfrüchte wie Ananas, Mamao (Papaia-Frucht), Guyaven, Süßkartoffeln, Erdnüsse, Passionsfrucht (maracujá), Pergamotten, Abacate, Orangen, Grapsfrucht, Hauptnutzpflanzen dar. Sie werden zur Erntezeit an Bahn- und Busstationen reichlich feilgeboten.

Das Agrarökosystem Bananal zeigt einen für das tropische Paraná und den ganzen Süden Brasiliens charakteristischen Jahreszyklus. Januar: Bodenbearbeitung für Kartoffeln, Getreide und Gemüse. Verschiedene Gemüsearten wie Kohl, Salat, Spinat, Knoblach, Zwiebeln usw. werden gesät. Andererseits beginnt die Ernte für Ananas, Bananen für Eigenbedarf, Paprika und Papaia-Früchte. Seit ca. 10 Jahren hat dieses Ökosystem sehr unter Gradationen der afrikanischen Biene zu leiden. *Apis mellifera adansonii.* 1956 als Paar nach São Paulo gebracht, weist sie gegenwärtig einen Expansionsdruck von nahezu 200 km/Jahr auf. Bisher wurden in Paraná mehr als 100 Haustiere und ca. 10 Menschen getötet und zahlreiche schwer verwundet. Manche müssen noch Jahre nach dem Unfall wegen Nierenleiden den Arzt aufsuchen. Im Bananal, wie auch in den anderen Ökosystemen, gibt es kaum einen Bienenstock, der nicht von "Afrikanern" besetzt wäre. Dank ihrer großen Agressivität und ökologischen Valenz, haben sie die hier seit mehr als 150 Jahren arbeitende Biene italienischer Herkunft, *Apis mellifera*, verdrängt. Während der Schwärm-Monate bekommt die Feuerwehr fast täglich Alarm, um mit Flammenwerfern usw. die angreifenden Bienenschwärme zu bekämpfen.

Kreuzungsversuche beider Rassen und alle möglichen Anpassungsverfahren zur Zähmung der Afrikaner sind überall im Gange, können aber die katastrophenartige Lage mancher Bienenzüchter nicht so ohne weiteres überwinden. Brasilien mußte als traditionelles Honigexportland zum Import übergehen.

Die Viehzucht beschränkt sich auf Schweine und Rinder. Obgleich Kulturweiden das ganze Jahr zur Verfügung stehen, ist die Rinderhaltung doch nur unbedeutend. Als Milch- und Fleischvieh findet man zahlreiche Kreuzungen von europäischen Rassen mit Zebu-Stämmen. Während der lezten Jahre immer häufiger durch-geführte Kreuzungen zwischen dem physiologisch resistenten Zebu-Rind und dem Fleischproduzenten Charolais führten zu einer wesentlichen Aufwertung des Viehbestandes Ostparanás. Als Weidegräser finden wir hier *Brachiaria*-Arten im Gemisch mit anderen wertvollen Gräsern wie *Panicum maximum*, dem Elephantengras, sowie *Hyparrhenia rufa*, dem minderwertigeren *Stylosanthes gracilis*.

12. Tropischer Regenwald (Epiphytal u. Phytothelmal)

Dank dem warmen Brasilstrom kommt echter tropischer Regenwald bis zu 28° südlicher Breiten vor. Allerdings ist dieses außerordentlich charakteristische Ökosystem rein auf die Osthänge der den Küstenstreifen mehr oder weniger folgenden Serren beschränkt, kontrastierend zum subtropischen Araukarienwald der Hochebenen. Die Serren bilden eine multiple Scheidewand zwischen beiden Ökosystemen. Wir benennen daher genauer dieses Marginalsystem als Randtropischen Küstenregenwald. Biozönotisch und zoogeographisch handelt es sich um eine der interessantesten Isolationsgebiete der Erde (MÜLLER 1973).

Nach dem Substrat, welches am Fuße der Serra do Mar dominiert und das Bett der zahlreichen Vorfluter füllt, zu urteilen, können wir ohne weiteres zwei Grundtypen unterscheiden, zwischen welchen selbstverständlich Übergänge bestehen: Einmal das aus groß und kleinen Felsblöcken bestehende Rhithropsephal (ILLIES 1971), welches die Erosionsschluchten und Vorserrengürtel säumt. Die mittlere Blockgröße nimmt mit zunehmendem Abstand von der Wasserscheide ab. Ihr Durchmesser ist in erster Linie der Flußgeschwindigkeit direkt proportional.

Diese Eigenart der Gebirgsflüsse, ihre Schotter überraschend genau zu sortieren, möchte ich granulometrische Potenz nennen. Sie kommt gerade in dem hiesigen tropischen Randgebirge, welches aus ziemlich einheitlichem und resistentem geologischem Material besteht, paradebeispielhaft zur Geltung. Das Oberteil des Systems des Rio Nhundiaquara mit seinen Nebensflüssen Màe Catira, Sào Joào Ipiranga, Conceiçào, Marumbi Pinto Passa Sete und Sambaqui bieten ausgezeichnete Beispiele hierfür. Der Rio Ipiranga konnte bereits den Kamm der Serra rückschreitend schneiden, so daß sein Quellgebiet jetzt im Hochland liegt, wobei mehrere an und für sich in das Iguaçú — System fließende Bäche nach Anzapfung jetzt in den Atlantik fließen.

Zweitens läßt sich in Richtung der eigentlichen Kulturebene hin das sog. Potamopsammal unterscheiden, welches überall dort zu finden ist, wo die Atlantikflüsse ihren reißenden Gebirgsflußcharakter verloren haben. Anfangs, so z.B. im Kessel von Porto de Cima ist das Psammal reichlich mit Schotter und kleineren Felsblöcken vermischt, welches mit zunehmender Mäandrierung den eigentlichen Flußsandbetten Platz macht. An der Nhundiaquarabrücke von Porto de Cima wurde von mir erst kürzlich eine sehr interessante Mesopsammalzönose entdeckt, welche ich nach der dominanten Art *Forficatocaris*-Zönose nannte. Das Genus *Forticatocaris* ist bisher nur im Grundwasser echt neotropischen Mesopsammals gefunden worden und scheint auch wirklich auf die Warmzonen Südamerikas beschränkt zu sein. Die für das paranänser Randtropengebiet charakteristische Art ist *Forticatocaris noodti*, welche sich von den im tropischen Teil des Amazonas und Paraná-Becken vorkommenden Arten durch sehr gute Merkmale unterscheidet.

Der Baum-Artenreichtum des Tropenwaldes ist auch dem paranänser Küstenrandtropenwald (= *Hygrodrymium marginale*) eigen. Immergrüne feuchtwarme Vegetation mit ausladender Kronenbildung in verschiedenen Schichtlagen garantieren lichtarme, epiphytenreiche und breitblättrige lichtreflektierende Urwaldformationen. Während der Hauptregenzeit steigt die Niederschlagshöhe auf über 300 mm pro Monat (Februar) mit zahlreichen Tagen von über $30°C$ im Schatten. Obgleich heute noch der Regenwaldkomplex in seiner primordialen Holzmasse weiter besteht und einen nur unwesentlich angebrochenen Gürtel darstellt, wird er doch in nächster Zukunft trotz warnender Stimmen kaum dem mit modernsten Maschinen arbeitenden allgemeinen Rodungsdruck standhalten können. Viele guten Vorhaben des seit ca. 5 Jahren wirkenden Brasilianischen Instituts für Forsten sind aus dem Planungsstadium noch nicht herausgekommen. Indessen geht man daran, in den von der Regierung abgegrenzten Schutzgebieten echte Kontrollarbeit zu leisten. Jeder Edelholzbaum soll auch nach seiner Wachstumseigenart als Individuum behandelt werden.
Von weit her heben sich aus dem Urwalddickicht mit Hauptkronenschicht die Baumriesen hervor, wie z.B. *Ficus* sp., *Schizolobium parahybum*, welche von den Einheimischen gefällt und lokal zur Kanu-Fabrikation verwendet werden. Hochgebaut sind auch noch folgende Bäume: *Piptadenia rigida, Andira anthelmithica, Phithecelobium* sp., *Melanoxylon brauna, Myrocarpus* sp., *Machaerium* sp., sowie *Jacarandá*-Arten, *Cassia*-Arten, *Enterobobium*-Arten, *Osmosia*Arten.

Die Bignonicaceen, wie *Tabebuia alba* und *Tabebuia umbellata* dienen mit *Jacarandá mimosaefolia* und *J. Semiserrata* vorzüglich zur Hartholzverwendung wie Schiffs- und Karosseriebau, sowie Bahnplankenlegung.

Die Lauraceen liefern zahlreiche Edelhölzer, wie z.b. *Nectandra nitídula, Cryptocarya moschata, Cryptocarya garuva, Persea racemosa* und die an Wert gegenwärtig sehr zunehmende *Ocotea pretiosa*, welche das devisenbringende

Sassafraz-Oel liefert. Anacardiaceen, wie *Astronium frax*, dienen hauptsächlich zur Möbelfabrikation sowie dem Schiffsbau. Weniger wirtschaftlich wichtig sind Vertreter der Apocinaceen, wie *Aspidosperma australe, Bombacaceen*, wie *Bombax cyathophorum* und Melastomaceen, wie *Tibouchina pulchra*, sowie einigen Myrtaceen-Arten: *Psidium cattlcianum, Marlieria gigantea, Gomidesia spectabilis, Myrtia publipetala* etc.

Die Hauptmittelkronenschicht weist folgende Charakterarten auf: *Cecropia adenopus*, welche bezeichnenderweise die von FRITZ MÜLLER im vorigen Jahrhundert entdeckte Ameisen-Coccidien-Symbiose aufweist, welche die jungen Bäume gegen die Blattschneiderameisen erfolgreich verteidigen. Dann *Lobelia langeana* und *Euterpe edulis*, sowie *Arecastrum romanzoffianum* und Baumbam-busarten, *Chusqueia* sp. Reicher Araceenwuchs (*Gleicheneria* sp. und *Phyllodendron*) sowie *Bromeliaceengalerien* (pine-apple-Familie) mit duftenden Orchideen, Kletter- und Schlingpflanzen tragen wesentlich zur Verdichtung der schwamm-feuchten, mit Flechten, Moos und Tillandsia belegten relativ dunklen Stamm-schicht bei. *Passiflora edulis* (= Passionfrucht = Maracujá), welche jüngst auch in Afrika angebaut wird, gehört zu den einheimischen Rankenpflanzen des brasilianischen Tropenwaldes.

Die Randlage des südbrasilianischen Dschungels kommt auch durch die Zusammensetzung seiner Fauna zum Ausdruck. Zahlreiche endemische Arten sind hier zu finden. Obgleich eine Bestandaufnahme noch immer aussteht, kann man doch sagen, daß unter den Wirbeltieren die Vögel und Schlangen auch hier am häufigsten vertreten sind (MÜLLER 1973). Unter den Wirbellosen sind es zweifellos die Insekten, von denen insbesondere die Apoidea, die Lepidoptera und Cerambycidea von den paranänser Zoologen wissenschaftlich bearbeitet werden. Zahlreiche Vogel- und Säugerarten bilden hier geographische Rassen, deren nächste Verwandte in disjungierten Populationen in der Hylaea des Amazonasbecken vorkommen.

Während die systemtypischen Großäuger wie Jaguar, Nabelschwein, Ameisenbär, Kapuziner-Affen weitgehend dem Jagdfieber zum Opfer gefallen und praktisch zum Aussterben verurteilt sind, konnten sich typische Rep-tilien- und Amphibienarten, welche substratgebunden sind, in den noch unberührten Urwaldrestgebieten erhalten. Von den camäleonähnlichen Iguaniden, die außer in der Neotropis nur noch auf Madagaskar und den Fidji-Inseln vorkommen, finden wir hier Anolis chrysolepis.

Die Geckoniden sind ebenfalls mit mehreren typischen Arten vertreten. Auch die heißen Feuchtluftwäider Ostparanás stellen für mehrere Tierfamilien optimale Bedingungen für eine außerordentliche Entfaltung dar. Allerdings steht auch hier der allgemeinen Regel folgend, eine ausgesprochenen Individuenarmut dem allgemeinen Artenreichtum gegenüber. Man muß selbst für viele Insekten spezielle Jagd- und Fangmethoden anwenden (Lockstoffe, Quecksilberlampe) und die nötige Geduld und Ausdauer aufbringen, um die Entomofauna unseres tropischen ostparanänser Randurwaldes einigermaßen erfassen zu können. Der Mangel an Humusbildung läßt die Vegetationsschicht vielerorts direkt auf die mehr als 20 m hohe Letten- und Sandschicht aufsitzen. Die Urwaldflüsse sind also auch hier wie in den echten weiten Tropen auch relativ arm an gelösten Mineralstoffen, im Gegensatz zu den kolloiden, hauptsächlich aus Eisen- und Aluminiumhydroxyden bestehenden Gewässer der Regenzeit. Andererseits sorgen die häufigen Platzre-gen, verbunden mit den Tages-Nachtwinden, stets für das Einfallen von organischer

Substanz, sei es in Form von Blättern, Insekten aller Art, Detritus usw., welche eine wertvolle Nahrungsgrundlage für die höhere Wasserfauna darstellen und indirekt das allgemeine Fehlen von Mineralsubstanzen wenigstens teilweise kompensieren.

Dank des Treibhausklimas im Urwaldinnern laufen sämtliche Stoffwechselprozesse mehr als doppelt so schnell ab wie in gemäßigten Breiten, so daß weder Zeit zur Bildung geeigneter humusdruchdrungener Substrate, noch deren typischer in den kälteren Breiten häufigen Microfauna der Bodenschicht besteht.

Es fehlt also praktisch die Bodenbiomasse interstitieller Grundsysteme, welche für das Massenvorkommen ökotypischer höherer Tiere Voraussetzung ist. Es überrascht daher auch weiter nicht, daß ganz allgemein im tropischen Urwald die Produktion an tierlicher Masse als relativ gering zu veranschlagen ist. Massenvorkommen von Urwaldsäugern und anderen Großtieren sind mit wenigen Ausnahmen (Affen- und Schweinerudel) sehr selten. Auch der Uwaldmensch hat unter dieser Biomassenlücke zu leiden. Er sucht sie durch Methoden, wie z.B. Brandrodung und Hackbau zu überbrücken. Auch heute noch zwingt dieses Urwaldmineraldefizit den primitiven Siedler von Zeit zu Zeit seinen Standort zu wechseln, also eine Art Halbnomadentum zu leben, ganz abgesehen von der erdrückenden Zahl von Schädlingen. Begriffe wie Bodenständigkeit und Scholle sind hier völlig fremd, ganz abgesehen von dem ständig wachsenden Sog, welcher durch die rapid evoluierenden Städte ausgeübt wird. Wer lesen und schreiben kann, bleibt kaum mehr im Urwald, selbst wenn er sich in der Stadt mit einer sehr bescheidenen Stelle abfinden muß.

Auch der prähistorische Brasilianer, also die Tupi-Indianer, lebten vorwiegend außerhalb des Urwaldes als jagende und fischende Nomaden. Die zahlreichen Meeres-Muschelschalen-Depots, sog. Sambaquis sind Zeugen ihrer Hauptnahrung und Siedlungseigenheiten. Sambaquis gibt es aber nur in Buchtnähe und auf den Inseln, während verhältnismäßig große Schnecken (*Strophocheilus* sp.) auch heute noch im Urwalddickicht zahlreich sind (Delikatesse wie *Helix pomatia*) und gelegentlich von Sammlern verspeist werden. Aus dem hiesigen Ökosystem sind jedoch keinerlei Sambaquiähnliche prähistorische Kulturenzeugen bekannt.

Das gefürchteste Großsäugetier ist zweifellos der Jaguar. Nach den letzten Informationen kann er noch in der Gegend von Guaraqueçaba, Serra Negra, also im Nordteil des vorliegenden Ökosystems, erbeutet werden.

Fruchtfressende Fledermäuse und Vampire (*Desmodus*) machen dem Waldbaurern mit seinen Vieh immer noch viel zu schaffen. Der Flachlandtapir (*Tapirus terrestris*) ist in Paraná praktisch verschwunden.

Der endemische Charakter des pluviotropikalen ostparanänser Regenwaldes kommt auch durch die Vogelwelt zum Ausdruck. Zahlreiche Lokalvarietäten (dank der ökologisch, orographischen Isolierung mit relativ hohem Spezialisierungs- grad) unter Vorkommen von Reliktformen, machen das Studium dieser Populationen ganz besonders interessant. Die Fortpflanzungszeit der Vögel des tropischen Regenwaldes wird weniger durch den jährlichen Temperaturverlauf als vielmehr durch das charakteristische Niederschalgsmaximum bestimmt. Viele Urwaldvögel beginnen anfangs Januar zu Brüten, also zu Beginn der Regenzeit, welche nicht nur das Waldesinnere immerfeucht, sondern die Epiphytentrichter stets reich mit Süßwasser füllt, selbst auf den vegetationsarmen Buchtinseln. Vogelpopulationen haben im warmfeuchten Urwald der Küstenseite des Atlantikbeckens eine andere Brutzeit als die gleichen Artgenossen, welche die subtropischen, wesentlich trockeneren Westhänge der gleichen Serra do Mar bewohnen. Biotopdifferenzen verbunden mit Klimaunterschieden können also in gleicher geographischer Breitenlage zu einer beachtlichen Abänderung des Sexualzyklus führen. Diese physioökologische Isolierung geht mit der Varietätenbildung Hand in Hand.

Formationsstratigraphisch sind (gemäß der Pflanzendecke) die Substratschichten für das Nisten von grundlegender Bedeutung.

1. Die Bodendecke: Ganz abgesegen von Nestbauart und Hauptlebenstätigkeit einschließlich des Nahrungserwerbes zeichnen sich diese tropischen Bodenvögel (z.B. *Tinamus* sp.) noch durch andere Eigenheiten, so z.B. der Polyandrie aus. Insbesondere das Rhithropsephal am Fuße der Gebirgswand mit seinem charakteristischen Felsblocklabyrinth, bietet auch der Vogelwelt eine besonders feste und nischenreiche Unterlage für Nisten und Nahrung. Angesichts der festen Substratunterlage konnte der evolutive Spezialisierungsprozess in anderer Richtung ganz besonders vorankommen, so z. B. hinsichtlich der Nahrung, wobei wir z. B. innerhalb der Insektivoren Vögel ohne weiteres Ameisenspezialisten, Mückenspezialisten und Insektenlarvenfresser unterscheiden können, ähnlich also den Spechten, welche sich im Nahrungserwerb auf die Baumrindenfauna spezialisierten. Der Nestbautyp ist meist ein Schalennest, seltener ein als Spaltennest anzusprechender Bau.

2. Parallel zu den Bodennistern lassen sich wenigstens noch die heterogenen Populationen der Mittelschichtnister sowie der Kronennister unterscheiden. Hier steht noch ein weites Untersuchungsfeld offen für den Ornithologen, Ökologen und Tiersoziologen. Grundlegende Fragen, z.B. ob der Rabengeier (*Coragyps atratus*) seine Beute ausschließlich aufgrund seines ausgezeichneten Geruchssinnes findet, stehen auch heute noch offen.

Unter den zahlreichen Amphibien des Urwaldes finden wir überall Bufo ictericus, die Agakröte, welche euryök ist und ihr Bufotoxin enthaltene 'Milch' lediglich auf Druck hin spritzt. Zahlreiche Hyliden bewohnen die dunkle und feuchte Kronenschicht, wobei die zahlreichen Epiphytentrichter, hauptsachlich die der Bromeliaceen, als beliebte Laichstätten dienen. Während die individuenreichen aber verhältnismäßig artenarmen Echsen reine Giftträger sind, kommen auch im Urwald Ostparanas zahlreiche Arten von äußerst gefährlichen Giftschlangen vor (*Bothrops jajaraca, Bothrops jararacussu, Bothrops cotiara, Bothrops alternatus*). Erst während der letzten Jahre konnte der öffentliche Gesundheitsdienst auch in die entlegensten Siedlungen des ostparanänser Urwaldes seinen Wirkungskreis ausdehnen. Auch der einfachste Waldbauer weiß, wo er im Unglückfalle das nötige Serum zur Neutralisierung eines eingespritzten Schlangengiftes finden kann. Am häufigsten ist auch hier *Bothrops jajaraca* und zahlreiche Elapiden (*Micrurus*).

Da Brasilien das größte Tropenland der Erde ist, die Gifttiere aber ihre eigentliche Heimat in den Tropen haben, ist es nicht weiter verwunderlich, daß Brasilien wohl eines der an Gifttieren reichsten Länder ist. Neben den artenreichen Schlangen sich auch die Spinnen beachtliche Giftproduzenten und das Institut Butantan liefert spezifische Spinnensera, welche schon manches Menschenleben retteten. Dank der unermüdlichen Forschungsarbeit Prof. Dr. WOLFGANG BUECHERLS sind unsere Spinnen sehr gut bekannt. Als gefährlichste Tropenspinne gilt auch hier die schwarze Witwe (*Latrodectes lactans*).

Von den sechs Skorpionenfamilien, die in Brasilien vorkommen, sind die

Buthidae am giftigsten. Die meisten Unfälle werden durch die Arten *Tityus bahiensis* (schwarzer Skorpion) und *Tityus gerrulatus* (weißer Skorpion) hervorgerufen. Die Giftspinnen gehören zu den sog. Vogelspinnen (Aviculariden) bzw. den sog. Krebsspinnen (Mygaliden). Von letzteren kommen *Grammostoal, Lasiodora, Acanthoscurria, Pamphobeteus* als Nomadenspinnen nur in der Neotropis vor. Alle sind Raubnachtspinnen. Ihre Hauptnahrung besteht aus Insekten, Vögeln (Nesthockern), Ratten, Kröten, Schlangen. *Sericopelma fallax*, die schwarze Spinne, wird von vielen Waldbauern mehr gefürchtet als die bekannte südamerikanische Klapperschlange *Crotalus durissus*. Sehr interessant für ökologische Studien sind die ökotypischen Galeriespinnen.

Es ist beabsichtigt, innerhalb der ökologischen Kurse für Post-Graduierte eine eingehende zönologische Gesamtanalyse der Ostparanänser Urwaldbiotope durchzuführen.

13. Das Kataraktal (Subtropischer Regenwald - upper mountain rain forest)

In Höhenlagen von ca. 700 m, wo das Rhithropsephal Riesenfelsblöcke in den einschneidenden Schluchten birgt, findet der eigentliche tropische Regenwald Ostparanas seine Begrenzung.

Obgleich es an Feuchtigkeit nicht fehlt, fallen doch die Jahrestemperaturmittel angesichts der höheren Lage, was sich vor allem während der Wintermonate bemerkbar macht. Wasserreiche Fälle stürzen hier in die Erosionsschluchten, und kaum kann sich mehr eine geschlossenen Vegetations an den Steilhängen behaupten. Zahl und Eigenart der Wasserfallbildung ist derart bezeichnend, daß sie die ganze Landschaft und damit eine ökologisch gut begrenzte Einheit charakterisieren. Ich nenne daher dieses Ökosystem das Kataraktal, wobei ich gerade das System subtropischer Wasserfallbiotope herausstellen möchte. Es reicht in Paraná ungefähr bis hinauf zu 1400 m ü.M., wobei selbstverständlich das Wasservolumen abnimmt. Es handelt sich geomorphologisch um ein ständig unter 'Verjüngung' stehendes Ökosystem, welches dabei ist, immer mehr Quellflüsse des Hochlandes in seinen Einzugsbereich durch rückschreitende Erosion zu gewinnen. Das fallende Wasser übersättigt sich mit Sauerstoff und zersprüht auf der Felsblockoberfläche seiner Schlucht. Als Folge hiervon finden wir die Schluchtenluft stets mit O_2 – reichem Wasserdampf übersättigt auch bei relativ trockenem Wetter im nahen tiefliegenderen Urwald oder Küstenstreifen. Selbst die periodischen Tages-Nachtwinde können diese immerfeuchten Schluchten des Kataraktals nicht wesentlich beeinflussen.

Tiefe Wolkenschwaden hängen meist bis zur Mittagszeit in den Spalten, was die Lichtverhältnisse (der an und für sich schon sehr dürftigen Insolationsverhältnisse) sehr beeinträchtigt. Dieses 'Waschküchenklima' garantiert die Entwicklung ausgesprochen hygrophiler Pflanzen und Bodentiere. Die immerfeuchten Felswände sind reichlich mit Moosen und Algen belegt. Eine reiche Kleinfauna, hauptsächlich aus Nematoden, Harpacticoiden und Protozoen, Milben und Rotatorien zusammengezetzt, bevölkert dieses als interstitielles Limnophytal anzusprechendes Bodensystem.

Am frühen Nachmittag, wenn nach einem sonnigen Morgen die Hauptnebelschwaden verdunsten konnten, kommt Schatten in den Schluchtwinkeln auf, deren Ausdehnung bald den ganzen Schluchtenkessel einnimmt. Da das Leben der meisten Pflanzen und Tiere photoperiodischen Bedingungen unterworfen ist (Lang-Kurz-Rhythmus), sind diese Erscheinungen ökologisch von hoher Bedeutung. Rasch wird es Abend, hauptsächlich im winterlichen Kataraktal. Bei Vollmond und klarem Wetter bilden sich oft ausgesprochene, herrliche Landschaftsbilder, welche demjenigen, der gegen Abend an den Strand fährt, ein wahrhaftes Naturschauspiel bieten kann. Im Westen die glutrote Wand der Strahlen der untergehenden Sonne, welche noch weit ausholende Bündel zwischen die Zacken der Serra do Mar wirft, mit all ihrem Lichtspiel, im Osten aber schon der aufgehende Mond, welcher mächtig über der Küste steht und das Sonnenlicht vom Ozean her auf die Küste und Wellen bescheiden beleuchtet. Die bündelnde Wirkung der Serra do Mar läßt den Vollmondrand ganz besonders kontrastreich erscheinen. Aus dem Unterholz heben sich die nischenreichen Baumfarne hervor, wie z. B. *Dicksonia sellowiana*, sowie die Cyathaceen *Alsophila schauschin*, *Cyathea caesarina* und *Hemitelia ripara*. Hier trifft man ab ca. 500 ü.M. die für das südneotropische Hochland charakteristische *Araucaria angustifolia* zusammen mit den Palmen *Arecastrum romanzoffianum*, *Bactris lindmaniana* und *Geonoma elegans*, deren Blätter als Ziegelersatz zum Abdecken der Urwaldhütten benutzt werden. Wie im rein tropischen Tieflandsurwald sind hier auch die Kletterpflanzen sehr häufig (Asclepiadaceen, Bignoniaceen, Malpighiaceen, Sapindaceen, Verbenaceen). Bambusstauden (Chusqueia-Arten) sorgen für oft nur mit dem Buschmesser zu lichtendes Unterholz. Luftwurzeln der Araceen hängen allerorts von den Baumstämmen herab. Die Menge der Epiphyten ist außerordentlich zahlreich. Hauptsächlich sind es wieder Bromeliaceen (*Vriesia* sp., *Nidularia* sp., *Aechmea* sp., *Tillandsia usnoides* und *T. tenuifolia*). Diesen reihen sich zahlreiche Arten aus der Familie der Polypodiaceen, Lycopodiaceen und Hymenophyllaceen an. Im Walddickicht sind herrliche Orchideen zu finden, welche charakteristischerweise hauptsächlich den Genera *Epidendrum*, *Masdevallia*, *Maxillaria*, *Neolanchea*, *Oncidium*, *Pleurothallis*, *Promenaea* und *Octomeria* angehören.

14. Das Nannobambusial

Mit zunehmender Höhe werden auch hier Baumvegetation, Wasserfälle und Schluchten immer kleiner bis sie ab ca. 1200 m einem eigenen Ökosystem Platz machen, welches wir als Nannobambusial bezeichnen. Auf dürftigem Boden, wobei auch frequent Bromeliaceen teilnehmen, kommt als Charakterpflanze der Zwergbambus (*Chusquea pinnifolia*) vor. Es handelt sich um eine Buschformation, welche weiter aufwärts ab ca. 1400 m ü.M. mehr oder weniger große Bergwiesen einschließt. Dieses an die Alm europäischer Prägung erinnernde Hochwiesensystem ähnert mancherorts sehr der Savanne mit eingesprengten Büschen aus Kompositen, Ericaceen und Melostomataceen u.s.w. Diese eigenartige Mittelgebirgsformation neigt vielerorts zu Muldenbildung mit Grundwasserstauung am Quellhorizontgürtel unter Sumpf- und Moorbildungen, welche zur Regenzeit unweigerlich überlaufen. Die Bodenbromeliaceen, deren Trichter zur Regenzeit ökologisch beachtliche Wassermengen mit typischer Kleinfauna beherbergen, können als dichte Bestände auftreten und weithin den Landschaftscharakter der Höhenlandschaft prägen. Diese stacheligen, gastfeindlichen Substrate, werden von Mensch und Vierbeinern gemieden. Dagegen stellen sie ein willkommenes Biotop zum Nisten für viele Gebirgsvögel, Schlangen und Eidechsen dar. Die Artenverarmung ist hier allgemein und steht klar vor Augen. Sie bezieht sich auf die Orchideen, welche dies oft durch Massenblüten von gewissen Arten, wie z.B. *Sophronites coccinea* ausgleichen. Obgleich ein Vergleich mit analogen alpinen oder andinen Formationen sehr verführerisch erscheint, ist es sowohl vom botanischen als auch zoologischen Gesichtspunkt aus nicht angängig, direkte Beziehungen zwischen den Regionen herzustellen.

15. Subtropische Regenwälder

Sie würden das Spiegelbild zum Kataraktal darstellen, wenn die auf der Westseite der Serra do Mar herrschenden hydrologischen Verhältnisse gleich wären. Dies trifft jedoch keineswegs zu. Dieser Westhanggürtel der Serra leitet als eigenes Ökosystem vom Nannobambusial der Hochregion zum ca. 900 m ü.M. liegenden Gebirgsvorland mit den zahlreichen Hügeln, Bruchwiesen und Quellkopfwäldern des Iguaçu-Tales über. Diese Westhänge sind pluviometrisch im Westen wesentlich trockener.

Oft , ganz besonders während der Wintermonate, regnet es im Küstenvorland, während über diese subtropische Gegend der Westseite hell die Sonne scheint. Die trennende Wirkung der Serragipfel kommt auch hier wieder eindrucksvoll zur Geltung. Dies gilt selbstverständlich für alle anderen klimatischen Faktoren. Photoperiodisch ist hervorzuheben, daß die Hauptlichtmenge während der Nachmittagsstunden eingestrahlt wird. Im Winter legt sich oft kalte, aus dem Süden eingeströmte Kontinentalluft auf die Westhänge und 'verankert' sich im weiten Iguaçutal, um tagelang naßkalte Witterung herrschen zu lassen. Dichte Nebelschwaden hüllen dann sie Landschaft ein. Fröste sind dann auch bis in den August hinein durchaus möglich. Trotz dieser klimatischen Unterschiede finden wir, soweit wir schließen dürfen, doch auch hier sehr ähnliche floristische und faunistische Formationen. Zusammenhängende Wälder sind ebenfalls reich an Epiphyten, hauptsächlich Bromeliaceen, Araceen, Orchideen und Lianen. Bemerkenswert ist das relativ häufige Vorkommen von Xerophyten, wie z. B. *Cereus peruvianus*, welche als Relikte eines Pleistozän in Paraná herrschenden semiariden Klimas aufgefaßt werden können.

16. Das Iguaçú - Tal

Der Rio Iguaçú stellt die Hauptader des paranänser hydrographischen Systems dar. Dank der seit Ende des Tertiärs anhaltenden Hebung der Serra do Mar wurden immer mehr der ursprünglich in Richtung des Atlantik fließenden Quellflüsse in das Curitibaner Becken geleitet, wo sie z. Zt. stark gebremst und unter Mäanderbildung im Sammeltal des Iguaçú nach Süden bzw. später nach Westen abfließen; Im Iguçútal erlangt das Curitibaner Becken gleichzeitig seinen tiefsten Punkt, so daß die Zubringerflüsse Atuba, Belém u.s.w. die Stadt Curitiba in NW-SO Richtung drainieren. Der Rio Belém, welcher Abwässer von Gerbereien, Schlächtereien, holzverarbeitender Industrien und chemischer Fabriken aufnimmt, kann als die natürliche Kloake der mit Hochhäusern gespickten paranänser Hauptstadt Curitiba, bezeichnet werden. Sein Flußbett muß ständig begradigt, verbreitert und ausgebaggert werden, damit er das rapid wachsende Angebot verschmutzter Gewässer in den Iguaçú bei São José dos Pinhais schütten kann. Trotz ausgiebiger Kanalisierungsarbeiten im Stadtbereich kommt es während der Hauptniederschlagsmonate zu unangenehmen Überschwemmungen im Stadtzentrum und in der Gegend der Mündung in den Iguacú, welche nicht selten die ganze Talaue des Iguaçú bis zur 1. Terrasse wochenlang unter Wasser halten können. Die periodischen Januar-Februar-Überschwemmungen reflektieren in der Zusammensetzung der Biozönosen.

Es fehlt an ausgedehnten, einheitlich, durchlässigen, grobkörnigen Sandschichten im Curitibaner Iguaçú-Tal, die als natürliche Langsamfilter zu einer gewissen Selbstreinigung beitragen könnten. Wo Sandablagerungen vorkommen, sind die meist mit undurchlässigem Lehm verschmutzt, so daß der Sand zu Bauzwecken erst gewaschen und gesiebt werden muß, bevor er Verwendung finden kann. Derartige Sandgewinnungsplätze bleiben dann als sog. cavas do Iguaçú, aufgegeben zurück und bilden eine Reihe von Teichen während der Regenzeit. Eine typische Kleinfauna in diesem Limnopsammal ist kaum vorhanden, da angesichts der ständigen Verstopfung der interstitiellen Kleinräume nicht das angebrachte Substrat hierfür gegeben ist. Die sog. cavas daher als Wassereinheiten zur Selbstreinigung der verschmutzten Belem-Wasser zu verwenden, erscheint daher von vorne herein als zweifelhaft, es sei denn, daß die entsprechenden technischen Vorkehrungen getroffen würden.

Seit ca. 10 Jahren begann die ursprünglich reiche Fischfauna des Iguaçú zu schwinden, und mit ihr ein Heer von Wasservögeln. Die Fischerei ist fast völlig verschwunden. Lediglich ein kleiner Detritusfresser (*Corydoras julii*) kann sich noch am Leben erhalten. Er muß hierbei aber zur Trockenzeit, die oft monatelang anhält, periodisch an die Oberfläche kommen, offenbar um sich mit Sauerstoff zu versorgen, der nach eigenen Analysen dann praktisch geschwunden ist. Auch hat er physiologisch die Möglichkeit, ähnlich den in Massen am Wasserrande vorkommenden Tubificiden, mehr Hämoglobin in seinem Blute zu konzentrieren, um seine Sauerstofftransportkapazität entsprechend zu erhöhen. Sein teilweise transparenter Körper erscheint dann blutrot.

Obgleich auch hier der Bekämpfung der Umweltverschmutzung politisch große Priorität eingeräumt wird, fehlen Kläranlagen (u.a. mit Methangasgewinnung als Nebenerwerb). Währenddessen greift aber durch progressive Infiltration die allgemeine Wasserverschmutzung auch auf das Grundwasser über, was besonders in subtropischen und tropischen Gegenden alamierend ist. Gegenwärtig stellt also das Ostparanänser Iguaçú-Tal, besonders im Curitibaner Becken, ein stark belastetes Ökosystem dar, gekennzeichnet durch wenige resistente Arten, welchen es gelingt, in diesem toxischen Milieu noch zu überleben.

17. Das Araucarial

Das hügelige Hochland (Planalto) war noch um die Jahrhundertwende mit ausgedehnten Araucarienwäldern bedeckt. Heute findet man nur noch Restbestände in den sog. Capões und Quellkopfgaleriewäldern welche die Flüsse begleiten zusammenhängende Bestände der *Araucaria angustifolia*. Eigentliche Araucariengroßwälder gibt es in Paraná nur noch im weiten Westen, wo sie rasch der Säge zum Opfer fallen. Erst in letzter Zeit geht man an ihre Afforstung. Es muß vermieden werden, daß die amerikanische Föhre *Pinus eliotii* die wertvollere, jedoch langsamer wachsende Araukarie ersetzt. Aber selbst die Restbestände der Quellkopf-galeriewälder fallen dem Menschen zum Opfer. Schon kann man sich in der Umgebung von Curitiba die Capões an den Fingern abzählen. Ausgedehnter, hügeliger Camp dominiert überall.

Das Araucarial ist somit völlig anders als der tropische Regenwald, wo die Araucaria nur gelegentlich als Fremdling gedeiht. Ihre klimatische Grenze ist um 500 m ü.M. zu suchen. Araucarien sind typischerweise mit *Podocarpus, Ilex, Cedrela*, Cyathaceen usw. vergesellschaftet und nur verhältnismäßig unbedeutend mit Epiphyten besetzt.

Auch fehlt es am Kletterpflanzendickicht des Urwaldes. Trotzdem ist es im Innern eines Araucarienwaldes schon alleine wegen der Schirmwirkung der Araucarienkronenschicht sehr dunkel. Auch der Unterholzbestand ist wesentlich reicher als wir ihn z. B. in einem deutschen Nadelwalde finden.

Nach einer floristischen Bestandsaufnahme des am Rande Curitibas unter Natur-schutz stehenden Capão de Imbuia zu schließen darf man folgende Zusammen-setzung für einen derartigen Araucarienrestwald annehmen: Bäume 6,8%, Epiphyten 10,6%, Sträucher 31.8%, Kräuter 22,2%, Gräser 26,5% und Kletterpflanzen 2,1%. Somit sind als Sträucher zu bezeichnende Pflanzenarten am häufigsten. Neben der Araukarie kommen als Bäume noch folgende Vertreter vor: *Ocotea porosa, (Lauraceae)*, Litraea brasiliensis *(Anacardiaceae)*, Schinus sp. *(Anacardia-ceae)*, Jacarandá puberula *(Bignoniacea)*, der Zimtbaum *Capsicodendron* (Canella-ceae), *Erythroxylum argentinum* (Erythroxylaceae), *Sebastiana brasiliensis* (Euphorbiacea), *Casearia decandra* (Flacourtiacea), *Xylosoma ciliatifolium, Brittoa glazumaefolia* (Myrtacea) und *Rhamnus sectipetala* (Rhamnaceae).

Zusammenfassung

Eine Intensivierung der tropenökologischen Forschung ist zwingend erforderlich: wertvolle, einmalige Ökosysteme des tropenbiotopreichsten Landes der Erde werden vom Menschen bedroht, bevor sie überhaupt wissenschaftlich bekannt sind. Es wird hier erstmals versucht, die Ökosysteme Ostparanás zu erfassen. Differen-zierungsreichtum und markanter Wechsel ökologischer Grundfaktoren auf leicht überschaubarem Raum machen die Arbeiten besonders reizvoll. Natürliche Landschaftsgliederung geht mit ökologischer Raumentwicklung Hand in Hand.

Die vorliegende Arbeit stellt einen Versuch zur Charakterisierung von 17 Makro-Ökosystemen in Großraum Ostparanás dar. Es werden (v. Ost n. West) unterschieden: 1. Kontinentalschelf, 2. Thalassopsammal (Strandlebensgemein-schaften), 3. Restingal und Dünenzone (Strandgehölzökotope), 4. Küstenhoch-wald (Epiphytal und Phytothelmal), 5. Tholomigal (Buchtweichböden), 6. Inselmangrovial, 7. Inselwald, 8. Inselpsammal und Inselpsephal, 9. Mangrovial, 10. Pantanal, 11. Bananal (Tropenkulturebene), 12. Der randtropische Küstenregen-wald, 13. Kataraktal, 14. Nannobambusial, 15. Subtropische Regenwälder, 16. Das Iguaçú-Tal, 17. Das Araukarial.

Ein datenreiches Großprofil ermöglicht die notwendige Gesamtschau der erarbeitenden Ökosysteme mit ihrem klimatischen, geomorphologischen und landschaftsökologischen Eigenheiten.

Resumo

Pesquisas em Ecologia Tropical deveriam estar em regime de urgência: ecosistemas dos mais antigos e de sumo valor produtivo para o Brasil, que é o país mais rico em biótopos tropicais, vivem sob ameaça de extinçao, antes de se tornarem conhecidos cientificamente. Este trabalho representa a 1ª tentativa de compreender os ecossistemas do Praná Leste sob o ponto de vista ecodinâmico. A riquesa em diferenciaçoes ambientais combinada com alteraçoes marcantes dos parâmetros ecológicos tornam as pesquisas especialmente interessantes, considerando ainda, que o espaço vital é de controle relativamente fácil. As divisoes naturais da regiao correspondem, diretamente ao desenvolvimento ecclógico do espaço vital.

Analisamos e caracterizamos 17 ecossistemas, na sua maioria ainda de ordem natural, preenchendo o macro-espaço Paraná – Leste: 1. A Plataforma Continental, 2. O. Talassopsamal, 3. O. Restingal e a Faixa de Dunas, 4. Epifital e Fitotelmal Costeiros. 5. O Tolomigal, 6. O Mangrovial Insular, 7. A Mata Insular, 8. Psamal e Psefal Insular, 9. O Manguezal, 10. O Pantanal, 11. O Bananal, 12. A Mata Pluviotropical Marginal, 13. O Cataratal, 14. Nanobambusial, 15. Matas Pluvio-Sub-Tropicais, 16. O Vale do Iguaçú, 17. O Araucarial.

Um perfil ecológico rico em detalhes possibilita pela primeira vez obter uma visão global e ampla da conjuntura dos ecossistemas, salientando suas caracrerísticas climáticas (diagramas), geomorfológicas e ecodinâmicas.

Literatur

Die vorliegenden Untersuchungen sind im wesentlichen Originalbeobachtungen. Sie erhielten Anregungen von folgenden grundlegenden Werken.

ABSABER & J.J. BIGARELLA 1961. Considerações sôbre a geomorphogenêse da serra do Mar no Paraná. *Bol. Paranaense de Geogr.* 4 e 5.

BIGARELLA, J.J. & R. SALAMUNI 1956. Planta provisória geológica da cidade de Curitiba e arredores, IBPT. Curitiba.

BIGARELLA, J.J. *et al.* 1969. Processes and environments of the Brazilian Quaternary. In: The Periglacial Environment. Montreal.

CABRERA, A. & YEPES, J. 1940. Mamíferos sudamericanos. Buenos Aires. Comp. Arg. Editores.

DANSEREAU, P. 1957. Biogeography, an ecological perspective. The Ronald Press. Comp., New York.

FITTKAU, J., J. ILLIES, H. KLINGE & G.H. SCHWABE 1968. Biogeography and Ecology in South America (2 vol.). Junk, The Hague.

HERTEL, R.J.G. 1950. Contribuição à ecologia da Flora, Epífita da Serra do Mar (vertente oeste) do Paraná. *Arq. Mus. Paranaense* VII.

HERTEL, R.J.G. 1959. Esboço fito-ecológico do Litoral Centro do Paraná. *Forma et Functio* I (6): 47-78.

HUECK, K. 1966. Die Wälder Südamerikas. Ökologie, Zusammensetzung und wirtschaftliche Bedeutung. G. Fischer, Stuttgart.

ILLIES J. 1971. Einführung in die Tiergeographie. Stuttgart.

JACOBI, H. 1953. Sôbre a distribuição da salinidade e do pH na baía de Guaratuba. *Arq. Mus. Paranaense* 10: 3-33.

JAKOBI, H. 1963. Sôbre a distribuição geográfica de Syncarida, *Dusenia* 8 (3): 115-125.

JAKOBI, H. 1974. Ecologia, (vol. 1). Wunderlich. Curitiba, Paraná.

MAACK, R. 1968. Geograficia Física do Estado do Paraná. UFP, Curitiba. Pr.

MÜLLER, P. 1970. Vertebratenfaunen brasilianischer Inseln als Indikatoren für glaziale und post-glaziale Vegetationsfluktuationen. *Abhdl. Dtsch. Zool. Ges. Würzburg* 1969.

MÜLLER, P. 1970. Durch den Menschen bedingte Arealveränderung brasilianischer Wirbeltiere. *Nat. u. Museum* 100 (1): 22-37.

MÜLLER, P. 1973. Centers of Dispersal of Terrestrial Vertebrates in the neotropical realm. Biogeographica 2: 1-248. Junk, The Hague.

ROSA, C.N. 1963. Animals de Nossas praias. São Paulo.

SCHIMPER, A.F.W. 1935. Pflanzengeographie, Jena.

SCHELFORD, V.E. 1963. The Ecology of North America, New York.

WETTSTEIN, R.v. 1970. Plantas do Brasil — Aspectos da Vegetação do Sul do Brasil. São Paulo.

Anschrift der Verfassers:

Prof. Dr. HANS JAKOBI, Universidade Federal do Paraná, Instituto de Biologia, Cx. P. 1923, Curitiba.

BIOPHYSICAL ASPECTS OF SPECIES DIVERSITY IN TROPICAL RAIN FOREST ECOSYSTEMS[1]

O. FRÄNZLE

Abstract

The exceedingly high species diversity of tropical rain forst synusiae on mature soils of the ferralsol and acrisol groups is attributed to the combined effect of very low nutrient content and high energy fluxes, which control the steady state of the macrosystem soils-vegetation.

The first part of the paper defines entropy of soils with regard to vegetation in terms of nutrient content and deals with the negentropic character of biocenoses under steady state conditions implying minimum production of entropy. Since in tropical lowlands the entropy of the impoverished mature soils is high, in terms of the above definition, the total entropy production of the corresponding rain forest synusiae must be proportionally low in order to maintain stability of the total macrosystem. This is accomplished by the evolution of hierarchically structured ecosystems of maximum diversity and a complicated arrangement in space. The last part of the paper deals with a numerical confirmation of the above hypothesis and is based on diversity analyses of rain forest assemblages in South America and Malaysia.

1. Introduction

It is one of the main aims of the earth sciences to develop models of the structure of the earth as a reaction sequence to different spatial and temporal inputs of matter and energy. The earth, and in particular its surface, the geosphere, is spatially differentiated in a number of more or less open complex systems. They show an historically-determined multiplicity and can be characterised in relation to systems theory according to their site, the numer and spatial ordering of their elements as well as their relationships (L.V. BERTALANFFY, 1971).

In the analysis of these systems, energy and energy transformations should be seen more than hitherto as the major framework within which biological as well as physical-geographical facts and theories can be considered, as part of a synthetic approach. As has been general for a long time with chemical and physical investigations, thermodynamics can also be emphasised as the basic accepted approach for both biological (see LEHNINGER, 1965; MARGALEF, 1968; EIGEN, 1971) and physical geographical problems (see LEOPOLD & LANGBEIN, 1962; FRÄNZLE, 1971). The following article may be considered as a contribution from this view point, to throw light upon the relationships between the soils of tropical lowlands and the long-recognised extraordinary floristic diversity[2] of their rain forests, after MÜLLER & SCHMITHÜSEN (1970) recently discussed 'Probleme der Genese südamerikanischer Biota'. The stimulus to enter this field of study came in particular from the papers on biogeography and ecology in South America (Monographiae biologicae 18, 1968 and 19, 1969) and the

---

1 Dedicated to the memory of Prof. Dr. h.c. CARL TROLL.
2 A definition of this in terms of structural parameters derived from species frequencies lists will be given in 2.1.1.

contributions by Fittkau and Brünig to the ecological symposium 1972 at the Max-Planck-Institut für Limnologie in Plön.

Since every climatic climax community presents a system of vegetation and soil components in dynamic equilibrium, the basic hypothesis in this article can be briefly stated: within given conditions the high energy flows of tropical rain forest climate combined with the low nutrient reserves of the climax soils give rise to maximum floristic diversity. Here, in contrast to the habitual inductive reasoning of natural scientists, an attempt will be made to validate this hypothesis (H) deductively, in order to lend it more precision for comparative inductive research. This can be done by ordering the hypothesis H in the more comprehensive (and therefore intuitively more elementary) theories T (thermodynamics of irreversible processess) and I (information theory) in which the basic concept of H (floristic diversity in the sense of McINTOSH 1967) is represented as a definable dimension of I and T. The hypothesis formulated above on basic relationships of rain forest ecosystems then becomes a testable statement of T or I, respectively.

2. Tropical rain forest as a system with self-regulatory organisation.[3]

The hypothesis H is linked logically with the well-known fact that tropical lowland rain forest is the most species-rich formation complex of the world. HUBER (after RICHARDS, 1952) reckoned the number of large-growing tree species as at least 2500. VAN STEENIS (1938) considered the number of phanerogam species of the Malaysian flora to be 20,000. In contrast the flora of central Africa, including some elements from the Savanna and montane forests, is relatively poor in species. According to ROBYNS (1946) the 170 phanerogam families of the Congo and Ruanda-Burundi are made up of 1631 genera with 9705 species. This richness in species of the rain forest was attributed by AUBREVILLE (1936) to a uniformity of demand by the species present and therefore, with little diversity in the environment, no one taxon 'winning' predominance. SCHMITHÜSEN (1959) traced the luxuriance, richness in species and form of the tropical lowland vegetation back to the fact that here, with constantly abundant moisture, more or less high temperatures predominate.

Little note has been taken in this context of the role played by site factors in the autecological sense, the generally extreme poverty of the tropical lowland soils, and to what extent the absence of one factor can be compensated by the overabundance of another. According SCHMITHÜSEN (op. cit.) in such compensations each factor necessary to life finds its limits at a definite minimum. Consideration of these questions will be within the sketched presentation as part of information theory and thermodynamic theory; floristic diversity of the lowland rain forest on poor soil can then be understood as a phenomenon of negative entropy.

2.1. Entropy and morphological differentiation

2.1.1. Diversity, entropy and information

An ecosystem can be presented as a canal transfering information (MARGALEF,

3 This term is applied sensu STEGMÜLLER (1970) and is not to be understood teleologically.

1961) with the distribution of individuals amongst the species as a parameter for canal width. If the numbers of representatives of the species a, b, c, ... s are designated Na, Nb, ... Ns, N being the total number of plants and if the probability that a plant belongs to species i is P_i ($P_i = N_i/N$) the average information content of a plant is

$$D = - \Sigma P_i . ld P_i \quad (1)$$

This is the definition of 'diversity' in information theory. It is attributable to SHANNON (1949) who employed it as a measure of information and designated it as 'entropy' (information e.). Diversity of an ecosystem or geosystem is more difficult to calculate by this formula than with the index proposed by McINTOSH (op. cit.).

$$I_D = \frac{N - \sqrt{\sum_{i=1}^{n} n_i^2}}{N - \sqrt{N}} \quad 4 \qquad (2)$$

but the SHANNON formula has the great advantage that it makes clearer the relationship between information entropy and thermodynamic entropy. The latter is the measure of the structural order of a system and in the formula

$$S = - k \Sigma P_i . ln P_i \qquad (3)$$

which corresponds exactly with formula (1), P_i is the probability that the system will be in quantum state i. It may again be objected that an equivalence of information or structural entropy and thermodynamic entropy is not acceptable. Nevertheless because of DENBIGH's (1959) work[4a] it can be argued that SHANNON described not only an analogy of form but a more deeply based relationship. FLECHTNER (1966) suggested that both cases are concerned with the fact that a system in position i in the next moment has a definite probability of being in position j, and that entropy is a measure of this probability.

The second principle of thermodynamics, that also applies to the relationship of an ecosystem to the environment, states that all physical processes proceed to achieve the greatest possible entropy in a system. The realisation of this tendency implies the exchange of heat of the system with the environment. Change in heat distribution and entropy have according to GIBBS the following relationship with free energy G

$$\Delta G = \Delta H - T\Delta S \quad (4)$$

ΔG is the change in the free energy of a system, ΔH the exchange of heat between the system and its environment, T the absolute temperature, and ΔS the change in entropy of the system. The free energy is thus the share of the total energy of a system that under isothermal conditions can perform work.

The particularly interesting performance of an ecosystem is the increase in its information extent, i.e. the increase in its (floristic) diversity. It ensues from

---

4 N is the total number of plants in the stand, n is the number of plants per species

4a DENBIGH (op. cit.) has shown that for a system with constant energy the entropy of the order is identical with thermodynamic entropy.

formula (4) that this occurs the more quickly the greater the free energy of the system is, and ΔG will diminish more slowly than entropy grows by up-take of heat from the environment. Further, energy exchange at higher temperatures leads to a greater increase of entropy than exchange at lower temperatures. This fact appears especially important in the present context since an ecosytem is not a closed system that can achieve a true thermodynamic equilibrium.

2.1.2 Open systems and dynamic equilibrium

The systems here discussed are thus open, and their equilibrium is therefore dynamic, 'all the forces at work on the system by opposed forces... are so balanced, that all the components of the system and their concentrations are stationary, although material flows through the system'. (LEHNINGER, 1970)

This dynamic equilibrium is the ordered state of an open system. All uniformly-running machines are for example characterised by this type of dynamic equilibrium. It follows from the thermodynamic theory of irreversible processes that in an open system entropy production is least for a definite energy transformation rate. As soon as the system is no longer in a steady-state condition it produces entropy more rapidly (stability criterium of GLANSDORFF & PRIGOGINE 1971).

A biocenosis in climatic climax conditions thus has a comparatively low entropy production and presents the final stage of a development that over sufficiently long periods could form ever more complex species, whose inner entropy has proportionally decreased. This seems to be in contradiction to the second principle that states every system strives to increase its entropy since the state of maximum entropy ist the most stable. In fact the above statement implies, however, that a system (or partial system) can lessen its entropy if that of the environment grows at a corresponding rate. Consequently the essential relationship may be formulated by:

rate of entropy production   = increase in entropy +
decrease in entropy   (5)

2.1.3 Morphological and functional differentiation as negentropic phenomena

For biology the essential conclusion follows that life is a 'constant struggle against the attempts[5] to produce, by irreversible processes, entropy' (LEHNINGER op. cit.) or as RIEDL (1972), following MARGALEF (1965, 1970) and MOROWITZ (1969, 1970), phrased less strongly 'it (the living system) overcomes to a certain extent the entropy principle by exporting more chaos than creating order.' This is achieved by means of synthesising ever larger and information-richer macromolecules[6], their grouping to form cell structures and by the evolution of more and more complex species, finally forming hierarchically structured biocenoses. This

5 The word 'attempt' is used as a mere 'façon de parler'. In fact the structure of 'goal-seeking' systems with the quality of self-organization can be characterised quite adequately without recourse to teleological terms (STEGMÜLLER op. cit.)

6 The individual bacteria cell of 2 μm diameter and 6.10^{-13} g weight has already an information content of about 10^{12} bits (LEHNINGER op. cit.)

impressive increase in order reduces the entropy in the living system, and compensates, or locally overcompensates, for the necessary (after the second principle) increase in entropy of the living system and its surroundings. Living beings and ecosystems behave negentropically: they produce, by preserving steady state conditions, entropy with minimum speed.

The stability criterion introduced above links the phenomenal conception of the evolutionary principle with thermodynamics. Darwin's principle appears as optimization principle based upon definite physical preconditions, and not as a basic irreducible phenomenon of the biosphere alone ... Concepts such as selection stress and selection value can be physically and objectively formulated by adoption of defined dynamic conditions (e.g. constant fluxes or 'forces') (EIGEN op. cit.).

2.2 The ecological differentiation of plant stands as a regulation process

The last point is not only of general but also of great special importance. It refers to the fact that in biological systems the degree of organisation and the spatial ordering of individuals are different reactions to the supply of soil nutrients and the tendency to maintain a stationary state. The effective relationship can generally be formulated as follows: the higher the entropy of the sub-system soil, as defined by the distribution of nutrients and trace elements, the lesser the rate of entropy production of the complementary sub-system vegetation must be, in order to realise the above mentioned minimum condition. Evolutionary history shows the following: Initially, there are relatively little-differentiated formations on soils with adequate to good nutrient supply and corresponding reserves of weatherable minerals. These systems are therefore in a relatively low state of entropy; they have only few realisation possibilities and are therefore improbable according to 2.1.1. which indicates that the number of realisation possibilities of a state is an expression of its thermodynamic probability.

If further development is considered unter the assumption that it is not influenced by external factors such as a climatic variation, it may be supposed that successive developmental stages of an ecosystem closely resemble each other. Since these sequential system states differ with respect to realisation possibilities, it is to be expected that the formations will develop towards stages that are more probable and thus show more realisation possibilities.

The most probable stage of the development under conditions of tropical lowlands will be an equal distribution of species in the stands, i.e. a macro-structure characterised by the greatest number of (geometrically) distinguishable realisation possibilities. This also entails a higher distribution entropy and is associated basically with the development of the soils and their litter layers in which the remaining macro- and micronutrients as well as trace elements become evenly distributed by transport processes and the activity of organisms. These adaptation and regulation processes which, as feedback phenomena, imply innumerable genotype changes, may be described as a 'learning process' in terms of information theory.

Here it is only comprehensible in behavioristic terms, i.e. in terms of external changes; the processes operative in individual systems of plants and leading to specific morphological and physiological constitution types[7] remain outside

7 Diversity analyses of part 3.2 would constitute considerably more essential statements if based on ecotypes rather than species lists. Ecotype lists for tropical rainforest, however, are not available.

consideration. All the described systems have thus the rank of 'grey boxes', the inner structure of which is little known, other than that certain stimuli (chemical and physical) are operative and produce responses.[8] The general problem is thus: what conceivable behaviour types of these 'boxes' exist?

In necessary limitation and in the light of the statements in 2.1. it can be said that because of high weathering speeds the rate of entropy production in moist tropical lowlands is high, leading to an even distribution of plants so that the macrostructural distribution entropy of the stands is also high. After 2.1.2 the stable state of an open system implies a dynamic equilibrium. Therefore the rapid increase of entropy in the soil must be balanced by corresponding negentropic processes in order to avoid the break-down of the system or at least to extend its existence. In this respect two reactions are particularly important: 1. Evolution of many new species and rapid immigration of existing ones to reduce entropy by the development of highly structured (from both morphological and taxonomic viewpoints) biocenoses; 2. Rapid re-cycling of nutrients from litter.

Both processes are energy-intensive, since structural order and high mineralisation rates both demand high energy fluxes. This fact forms the basis to formulate the above principles of plant distribution as minimum conditions.

Hence adaptative selection is most effective in the tropical lowlands because of the maximum radiation balances of these areas (see BUDYKO 1955), and because the fall of nutrient supply below the critical minimum (see below) is more effectively balanced than in the ectropics by distinctly higher re-cycling rates. In this connection it should be emphasized that the exceedingly high faunistic diversity of rainforest biota contributes to intensify the negentropic 'behavior' of the vegetation. If the energy flow as a parameter of the biocenotic dynamics is defined as the quotient of primary production / total biomass (MARGALEF op. cit.) it proves to be negatively correlated with the number of tropic levels. The entropy rate is consequently high in a poorly structured system, where bacteria play a large role in the decomposition of the primary production. It is clear that the regulation process achieves a balance, an adaption in the sense of Flechtner, with various grades ranging from optimum to just tolerable. If the critical threshold value is exceeded, because of inner or outer disturbances, regulation processes can no longer maintain equilibrium: the organism no longer 'adapts' to the new environment. Therefore the threshold values of individual factors in his ecological constitution must change and equilibrium must be re-established to meet the new environmental conditions. Only those organisms survive that are able to re-adjust themselves accordingly. The others die out or give way to concurrent organisms. In the framework of a more detailed analysis, various kinds of controlling and regulating mechanisms (auto- and external compensation, and a combination of both) may be distinguished (see STEGMULLER op. cit.; VOLTERRA 1926, LOTKA 1951, MARGALEF 1968).

The two most important natural disturbances of the stationary state of the tropical lowland rain forests are the Pleistocene climatic fluctuations (with the later outlined consequences for pedogenesis) and the loss of practically all the residual nutrients in comparison to acrisols and ferralsols, when normal tropical humic podsols (normal in horizon thickness) develop into giant podsols with more

---

8 Each reaction of a system, in the sense of information theory, is a message to the outer world.

than 1 metre E horizons. In 3.2 it is shown that rainforest stands on soils of this type and partly on histosols have an essentially reduced floristic diversity. This can be explained, if the cases so far investigated are representative, in terms of the stage in pedogenesis. On the sterile substratum the critical value for nutrient supply is not met; the re-cycling process is lengthened by the production of decomposer resistant litter, and only specialised plants can survive. Formations of this type may be found among the ericaceous variants of the Malaysian rain forest. Although it is in general impossible to calculate the entropy production of highly complex systems (see RICHTER 1972) it may be argued on the above-mentioned grounds that degraded stands of this type produce entropy more rapidly. They work 'uneconomically', their free energy and therefore their tendency to morphological (and consequently systematic) differentiation is less than that of stands on acrisols and ferralsols.

2.3. Hypothesis H as derived principle

The previous section showed, in the sense of the introductory basic relationship, that the biological notions of the present hypothesis are physically definable quantities. The term 'floristic diversity' appears as a particular deduction of information theory, and is determined, like the nutrient concentration of the soil, as a structural parameter that stands at any given time in unequivocal relationship to the entropy of the system. 'Concurrence', 'morphological and physiological constitution' as terms of causal botany in the sense of ELLENBERG (1956) can be interpreted as essential components of biological regulation processes. The fundamental principle of the present hypothesis, that provided sufficiently high energy fluxes and within certain boundary (i.e. stability) conditions the floristic diversity of the formation complexes of tropical lowland rain forest negatively correlates with the soil quality of the various sites, is thermodynamically substantiated by analysis of the basic relations of soil and vegetation as normal 'behaviour' of an open system. The (phenomenologically formulated) ecological principle of FRANZ (1952/3) 'The more continuously the milieu conditions of a site have developed, the longer similar environmental conditions have prevailed, the richer in species and the more balanced and stable is the corresponding biocenosis' appears physically deducible. At the same time the realm of validity of the first 'biocenotic law' (THIENEMANN 1918) 'The more variable the biotic conditions of a site the greater the number of species in the associated biocenosis' is more precisely defined.[9]

3. The inductive evidence of the hypothesis

With the derivation of hypothesis H from the two elementary theories T

[9] In a paper published after the preparation of the present article (FITTKAU (1973) hints at the possibility 'to explain the faunistic diversity of tropical rainforests in terms of specific adaption processes to the extremely differentiated vegetation cover' but he argues 'that the first biocenotic principle is not apt to yield a satisfactory explanation of the overwhelming floristic diversity of these forests.' Extending FRANZ' principle he considers the rainforest associations of Amazonia as 'biological filters' for 'the retention of allochthonous and autochthonous nutrients made available by the continuous de-composition of organic matter... Their filtering efficiency increases with the number of species present, which in turn contributes considerably to stabilize the whole macrosystem.

(thermodynamics of irreversible processes) and I (information theory) the lines of empirical evidence are indicated, and also object and scope of subsequent hypotheses definable. Testing hypothesis H implies a comparison of tropical lowland soils in which, perhaps with the exception of giant podsols, water plays no role as a minimum factor, with the diversity indices of different stands.

3.1 The soils of tropical lowland rainforest

3.1.1 The soils of the Hylaea

An summary view of the soils of lower Amazonia is given in the following table. The soils are ordered in terms of decreasing weathering intensity.

Table 1. The main soil types of lower Amazonia
orthic ferralsols[10]
xanthic ferralsols
plinthic ferralsols
orthic acrisols
plinthic acrisols
gleyic acrisols
humic podsols
pellic vertisols
eutric gleysols
plinthic gleysols
dystric gleysols
albic arenosols
ferralic arenosols
fluvisols

Most widely distributed in areal terms are the xanthic ferralsols,[11] developed on Plio-Pleistocene kaolinic freshwater sediments of the Barreiras Series in little dissected or non-dissected region. They form soil associations with plinthic ferralsols and acrisols[12] and inclusions of dystric and plinthic gleysols (cf. FAO-Unesco soil map, sheet IV, (1970). The more or less well developed terrace sequences of the terra firme areas, relatively close to streams, are pedologically characterised by the association xanthic ferralsols, plinthic ferralsols, plinthic and orthic acrisols including both the gleysols types, tropical lowland podsols (KLINGE 1968) and albic arenosols.

The Holocene floodplains are characterised by the soil association dystric gleysols, fluvisols, eutric and plinthic gleysols including gleyic and plinthic acrisols as well as xanthic ferralsols. Relatively rich in nutrients are only the fluvisols on the levées of the Solimões, the base rich sediments of which are probably derived from limestone areas of the Peruvian Andes. On the polycyclic planation surfaces spanning the old shields of Central Brazil and the Guayanas, ferralsols and orthic acrisols form the most extensive associations. They are often sandy and characterized by an extremely low exchange capacity of the clay fraction and extremely low base saturation.

10 Group names according to FAO-UNESCO classification (DUDAL 1968)
11 The cation exchange capacity of the B horizon in less than 16 mval/100g clay; the clay content is $>15\%$ of the fine earth fraction.
12 Acrisols are characterised by a base saturation of $<35\%$ in the argilluvic B horizon.

According to the above soil map the structure of the soil cover remains nearly the same up to the piedmont areas of the Peruvian Andes. But SIOLI (1968) had rightly emphasized that here the Plio-Pleistocene Pebas Formation is the substratum for soil formation, and these beds contain irregularly interbedded layers of brackish and even marine sediments. Since these are likely to stem from the erosion of the rising Andes they may contain more nutrient-producing minerals than the Barreiras sediments which originated for the major part from deep weathering mantles of the old granite and gneiss complexes of the Guayanas and Central Brazilian Shields. In favour of SIOLI's thesis it may be mentioned that the height and diameter of rain forest trees of the west Amazonian lowland are clearly greater than in central and east Amazonia (AUBREVILLE 1961). A further argument is the species composition of the várzea forests. In lower Amazonia they are an alien element in the riverine landscape: but nearer to the mineral-providing Andes the contrast is steadily weakened. 'With respect to vegetation the várzea forests of lower Amazonia, and the rain forest of the terra firme have scarcely any similarity, a phenomenon due to difference of the substratum, to different origin of the soil material, and therefore to different physico-chemical characteristics of the várzea and terra firme soils, in particular with reference to pH and nutrient content' (SIOLI op. cit.). Or as FITTKAU (1969) phrased it: the várzea of the area studied represents a funnel-like extension of upper Amazonian conditions; it is the result of intermittent transport of nutrient-rich Andean sediments by 'agua branca' streams and finally by the lower Amazon.

The many years of continued obervations necessary for a balancing investigation of the nutrient budget for a large area, involving precipitation and river-water chemistry (SIOLI 1956, 1964, 1968 a,b; GESSNER 1960, a, b, 1964, EDWARDS & THORNES 1970) are available only from the Rio Negro or the vicinity of Manaus (ANON 1972a,b), respectively. If the values from the latter locality are taken as representative of the catchment area of the Rio Negro then the phosphate and alkaline earth budgets seem balanced, iron shows a permanent removal out the geosystem, whilst chloride is enriched. This reveals immediately that the simple extrapolation on the basis of existing data is not reliable, since an effective accumulation of chloride is not possible under the climatic conditions of tropical rainforest. Further reference must be made to the fact that no clearer statement is available on the origin of substances in the rain-water. Corresponding research in the rainforest of the Congo (MEYER & DUPRIEZ 1959) have shown that linkages take place due to dust, leaves, etc. being blown aloft by gusts before the onset of rain, and then returned to earth after a short period (re-cycling). The low value of $KMnO_4$ consumption can be considered as an indication of the predominantly inorganic bonds of ions. If it is further considered that, as opposed to rainfall, much smaller quantities of total nitrogen are present in flowing water, a high need for this element is revealed (ANON 1972b). In fact the balance of the whole system vegetation-soil-water may be negative since the chemical characteristics of the Rio Negro, like other black-water streams, shows a strong depletion of Fe, Si, total and organic nitrogen, and a weak one of Ca, Mg, and P.

In the light of these findings the often quoted statement that the tropical rainforest with reference to nutrient supply represents a closed system requires revision. This is also established by the fact that acrisols belong to the most widely distributed soils of this formation complex. Under present climatic and vegetational conditions (and also during the Pleistocene, if the hygric fluctuations of this period did not exceed the threshold values of acrisol formation) they develop from less mature soil forms with better base status and clay mineral associations of distinctly higher cation exchange capacity. They ultimately evolve into ferralsols, and it is with these soils that further pedogenetic processes become slow, even on a geological scale.

An essential cause of the extreme nutrient deficiency of the Amazonian soils is the geomorphological evolution of this region, now known in basic outline, which appears to have been less differentiated than in other rainforest regions of the earth. In particular we refer to the investigations of BIGARELLA and ANDRADE (1965) limiting the last period of extensive planation processes stretching from Uruguay to the Amazon, to the older Quaternary. Observations in Brazil, and in particular in the peneplain areas of Africa, show that thereby soils are truncated and that the deeper and more nutrient-rich parts of the profile come to the surface. This results in a scatter of sites with improved soil conditions, whilst places that received the correlative sediments at least preserved their original soil quality. After this distinctly semi-arid period of planation the following climatic oscillations of minor significance (cf. DEMANGEOT 1959) only led to local or subregional stone-line development; soil truncation was distinctly less. Consequently the removal of nutrients took place at latest since the older Quaternary, possibly influenced by the minor fluctuations of hygro-climate. Also SCHWABE (1968), (following the views of RAWITSCHER 1946 and BOERGER 1948) has recently stressed the ecological importance of continual nutrient depletion.

3.1.2 The soils of the African rainforest

According to D'HOORE (1964) the soil cover of these regions is composed of rhodic and xanthic ferralsols, acrisols, gleysols, histosols, fluviosols, and lithosols,[13] in the eastern margins it consists of luvisols because of extensive, repeated truncation processes during the Quaternary. KLINGE (op. cit.) mentions groundwater podsols. Most widely distributed are the generally polygenetic ferralsols, the low fertility of which is dependent on the parent material and the clay content varying between 20% and 50%. Without fertiliser most degrade irreversibly after a few years of cultivation (cf. NYE & GREENLAND 1950). Irrespective of typological similarities African soils seem to have been more affected than the Amazonian ones by the morphological effects of the Pleistocene climatic oscillations (CLARK 1967, FAIRBRIDGE 1964, FRÄNZLE op. cit., DE HEINZELIN 1952, 1955, DE PLOEY 1965, VAN ZINDEREN BAKKER 1967). This presumably delayed the nutrient removal. In addition because of the lesser extent of the African rainforests (1.8×10^6 km^2) it should be noted that their soils must have been considerably more influenced by the morphodynamic processes affecting the often strongly accidented peripheral regions during the arid activity phases of the Pleistocene.

3.1.3. The soils of the Indonesian rainforest

The larger Indonesian islands, with the exception of Java where in particular young

13 Soil group correlation by author after FAO/UNESCO classification.

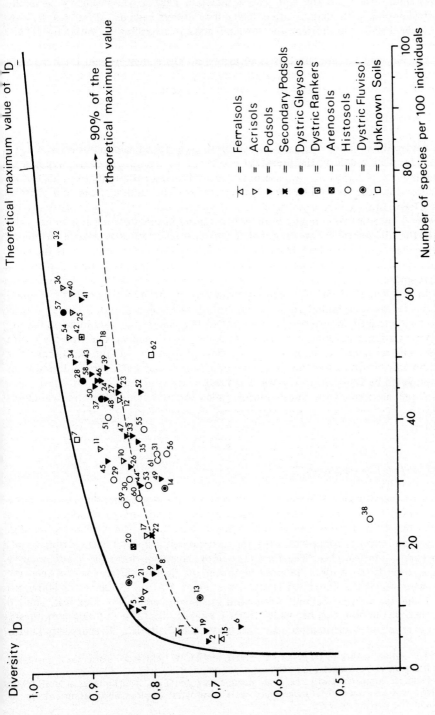

Fig. 1. Diversity indices of tropical rainforest associations.

Diversity I_D

Theoretical maximum value of I_D

90% of the theoretical maximum value

Number of species per 100 individuals

△ = Ferralsols
▽ = Acrisols
▶ = Podsols
✖ = Secondary Podsols
● = Dystric Gleysols
⊞ = Dystric Rankers
▣ = Arenosols
○ = Histosols
◉ = Dystric Fluvisol
□ = Unknown Soils

lavas and tuffs form the soil parent material, have a soil inventory generally corresponding with that of other lowland rainforest regions (cf. MOHR & VAN BAREN 1954). On Borneo the lowland podsols according to BRÜNIG (1957) cover more than 100,000 ha. Their base status is extraordinarily poor (particulary in Ca, Mg, and K) and the C/N ratio in raw humus is near 60. On these soils are heath forests, with as characteristic trees *Dacrydium elatum, Agathis borneensis, Castanopsis* as well as *Shorea stenoptera* and *Hopea* sp.

3.2 Diversity analyses of tropical lowland rainforest

A compilation of the diversity values of tropical forest communities computed from frequency data in the literature and BRÜNIG (lecture 1972) is given below. The indices of stands 30 - 62 are provided by Professor BRÜNIG from his lecture manuscript, the remainder was calculated by the author from the lists of AUBREVILLE 1961, DAVIS & RICHARDS 1933, 1934, and RICHARDS 1936.[14] In order to compare with similar figures in the literature it is to be noted that the abcissa values are standardized by introduction of a relative frequency parameter. This is derived from the equation N:100 = n:x, when N is the total number of plants in the stand being investigated, and n the total number of species represented.

In the sense of an inductive confirmation of the hypothesis H the diagram shows for most cases that on ferralsols, acrisols and tropical podsols the rainforests are characterized by diversity values near to the theoretical maximum. A sharper differentiation according to nutrient status and water budget is not possible with the data at hand. It is much to be desired that in the future comparative geobotanic investigations are supplemented with detailed soil profile descriptions as provided by BRÜNIG in Sarawak (lecture).

A second point is the unusually low index value of stand 38 on tropical moor, allowing the conjecture that its nutrient supply is below the critical value supposed above.

3.3 Further investigations

As a further working hypothesis it follows from the above, that on younger, more nutrient-rich soils in the Tropics less diversified stands are to be expected than upon ferralsols, acrisols, and podsols. Comparative investigations of várzea stand in lower and upper Amazonia with the occasionally adjacent terra firme stands appear to confirm this. 'The farther we move upstream form the lower Amazon west towards the Andes, the more disappear the floristic differences between the forest cover of the inundated lands of the upper Amazon (Solimões or Marañon) and the terra firma behind, until both are finally practically identical, and no further distinction can be made' (SIOLI 1968b). To fully test this hypothesis, however, exact floristic probes are necessary. A lesser floristic diversity should also

14 The author wishes to thank Miss G. GEHL and Mr. W. RENK for assistance in searching the literature and calculating indices.
To facilitate comparison with the literature an evaluation of the diversity indes of McINTOSH (1968) was employed. The undernamed work in progress employs information entropy to characterise floristic diversity.

be shown by stands on soils of the old paleozoic rock, stretching from the margins of the Barreiras formation.

Finally two further research schemes also based on hypothesis H are to be mentioned: 1. Comparative investigations of the floristic diversity of climax formations on similar soils in different climates; 2. Diversity analyses of climax stands on different soils in areas with comparable net radiation. Here particular attention is paid to the question within what limiting conditions nutrient status of soils and floristic diversity of the corresponding plant formations correlate negatively.

Zusammenfassung

Ausgehend von der Tatsache, daß eine klimatische Climaxgesellschaft mit ihren Böden ein stationäres Makrosystem hohen Ordnungsgrades darstellt, wird die Hypothese begründet, daß die geringen Nährstoffreserven der Climaxböden in Verbindung mit den hohen Energieflüssen des tropischen Tieflandsklimas die maximale floristische Diversität der Regenwälder bedingen.

Zunächst wird gezeigt, daß Climaxbestände im Vergleich zu ihrer Umwelt entropiearm sind, da Lebewesen und Lebensgemeinschaften sich negentropisch verhalten; sie erzeugen durch Aufrechterhaltung eines stationären Zustandes Entropie mit minimaler Geschwindigkeit. Da die Entropie der tropischen Climax-böden — charakterisiert durch ihre Nährstoffverteilung — hoch ist, muß die Entropiebildungsrate der auf ihnen stockenden Formations-Komplexe möglichst gering sein, um den stabilen Zustand des Gesamtsystems in der wirksamsten Weise aufrecht zu erhalten. Dies geschieht durch Ausbildung hierarchisch gegliederter hochgeordneter (diversifizierter) Bestände, die aus einer Vielzahl von Arten bzw. Ökotypen bestehen. Im letzten Teil des Artikels wird anhand einer größeren Anzahl von Diversitätsanalysen eine induktive Bestätigung der oben genannten Hypothese gegeben.

REFERENCES

ANON. 1972a. Die Ionenfracht des Rio Negro, Staat Amazonas, Brasilien, nach Untersuchungen von Dr. Harald Ungemach. *Amazoniana* III: 175-185.

ANON. 1972b. Regenwasseranalysen aus Zentralamazonien, ausgeführt in Manaus, Amazonas, Brasilien, von Dr. HARALD UNGEMACH. Amazoniana III: 186-198.

AUBRÉVILLE, A. 1936. La flore forestière de la Côte d'Ivoire. Paris.

AUBRÉVILLE, A. 1961. Etude écologique des principales formations végétales du Brésil. Nogent-sur-Marne.

BERTALANFFY, L. v., 1971. General System Theory. London.

BIGARELLA, J.J. & A.N. AB'SABER 1964. Paläogeographische und paläoklimatische Aspekte des Känozoikums in Südbrasilien, *z.f. Geomorph.* 8: 286-312.

BIGARELLA, J.J. & G.O. ANDRADE 1965. Contribution to the Study of the Brazilian Quaternary. *Geol. Soc. America, Spec. Paper* 84: 433-451.

BOERGER, A. 1948. La desmineralización de los continentes. Creciente desequilibrio productivo y nutritivo. *An. Soc. cient. argent.* 145: 281-306.

BRÜNIG E.F. 1957. Waldbau in Sarawak. *Allg. Forst- und Jagdztg.* 128: 156-165.

BUDYKO, M.I. 1955: Atlas teplovogo balansa. Glavn. Upr. Gidrometsluschby SSSR. Leningrad.

CLARK, J.D. 1967. The Atlas of African Prehistory. Chicago und London.

DAVIS, T.A.W. & P.W. RICHARDS 1933, 1934. The Vegetation of Moraballi Creek, British

Guiana: an Ecological Study of a Limited Area of Tropical Rainforest. Part. I. *J. of Ecology* 21: 350-384; Part II. *J. of Ecol.* 22: 106-155.

DEMANGEOT, M.J. 1959. Observations morphologiques en Amazonie. *Bull. Ass. Géogr. fr.* 286: 41-45.

DENBIGH, K. 1959. Prinzipien des chemischen Gleichgewichts. Darmstadt.

DUDAL, R. 1968. Definitions of soil units for the soil map of the world. World Soil Resources Reports 33 Rom.

DUDAL, R. & M. SOEPRAPTOHARDJO 1957. Soil classification in Indonesia. Pember. Balai Besar Penjel. Pertan 148.

EDWARDS, A.M.C. & J.B. THORNES 1970. Observations on the dissolved solids of the Casiquiare and Upper Orinoco, April-June, 1968. *Amazoniana* II: 245-256.

EIGEN, M. 1971. Selforganization of Matter and the Evolution of Biological Macromolecules. *Naturwissenschaften* : 465-523.

ELLENBERG, H. 1956. Grundlagen der Vegetationsgliederung. I. Teil: Aufgaben und Methoden der Vegetationskunde. Stuttgart.

FAIRBRIDGE, R.W. 1964. African Ice Age Aridity. In: NAIRN, A.E.M. (ed.) Problems in Palaeoclimatology 356-360. New York.

FAO-UNESCO 1971. Soil map of the world 1:5 000 000, Vol. IV, South America. Paris.

FITTKAU, E.J. 1973. Artenmannigfaltigkeit amazonischer Lebensräume aus ökologischer Sicht. *Amazoniana* 4 (3): 321-340.

FITTKAU, E.J. 1969. The Fauna of South America. In: Biogeography and Ecology in South America, Vol. 2, Monographiae Biologicae 19: 624-658.

FITTKAU, E.J., J. ILLIES, H. KLINGE, G.H. SCHWABE & H. SIOLI 1968, 1969. Biogeography and Ecology in South America. Monographiae Biologicae 18, 19. Junk, The Hague.

FLECHTNER, H.J. 1966. Grundbegriffe der Kybernetik. Stuttgart.

FRANZ, H. 1952/53. Dauer und Wandel der Lebensgemeinschaften. Schr. Ver. Verbr. naturw. Kenntnisse Wien. Ber. 93, Vereinsjahr 1952/53: 27-45.

FRÄNZLE, O. 1971. Physische Geographie als quantitative Landschaftsforschung. Schrift. Geogr. Inst. Univ. Kiel 37: 297-312.

FRÄNZLE, O. 1972. The late Pleistocene planation surfaces and associated deposits and soils of the Transvaal lowveld. In: International Geography 1972, Vol. 1: 251-253.

GESSNER, F. 1960a. Limnologische Untersuchungen am Zusammenfluss des Rio Negro und des Amazonas (Solimões). *Int. Rev. ges. Hydrobiol.* 45: 55-79.

GESSNER, F. 1960b. Ensayo de una comparación química entre el Río Amazonas, el Río Negro, y el Orinoco. *Acta cient. venez.* 2:3-4.

GESSNER, F. 1964. The limnology of tropical rivers. *Verh. intern. Ver. Limnol.* 15: 1090-1091.

GLANSDORFF, P. & I. PRIGOGINE 1971. Thermodynamic Theory of Structure, Stability and Fluctuations New York.

HEINZELIN, J. DE 1955. Observations sur la genèse des nappes de gravats dans les sols tropicaux. Publ. de L'INEAC, sér. sci. 64. Bruxelles.

HOORE, J.L. D' 1964. La carte des sols d'Afrique au 1:5 000 000. Publ. de la C.C.T.A. 93. Lagos.

KLINGE, H. 1968. Report on Tropical Podzols. FAO Technical Rep.

LEHNINGER, A.L. 1965. Bioenergetics. The Molecular Basis of Biological Energy Transformations. New York - Amsterdam.

LEHNINGER, A.L. 1970. Bioenergetik. Stuttgart.

LEOPOLD, L.B. & W.B. LANGBEIN 1962. The concept of entropy in landscape evolution. U.S. Geol. Survey Prof. Paper 500-A. Washington.

LOTKA, A.J. 1956. Elements of mathematical biology. New York - Dover.

MARGALEF, R. 1961. Communication of structure in planctonic populations. *Limnol. Oceanogr.* 6: 124-128.

MARGALEF, R. 1965. Ecological correlations and the relationship between primary productivity and community structure. *Mem. Ist. Ital. Idrob., suppl.* 18: 355-364.

MARGALEF, R. 1968. Perspectives in Ecological Theory. Chicago - London.

MCINTOSH, R.P. 1967. An index of diversity and the relation of certain concepts to diversity. *Ecology* 48: 392-404.

MEYER & DUPRIEZ 1959. zit. nach ANON. 1972b.

82

MOROWITZ, H. 1969. Energy Flow in Biology. Academ. Press. New York.

MOROWITZ, H. 1970. Entropy for Biologists. New York.

MOHR, E.J.C. & F.A. VAN BAREN 1954. Tropical soils. The Hague - Bandung.

MÜLLER, P. & J. SCHMITHÜSEN 1970. Probleme der Genese südamerikanischer Biota. In: Deutsche Geographische Forschung in der Welt von heute. (Gentz-Festschrift): 109-122.

NYEP.H. & D.J. GREENLAND 1960. The soil under shifting cultivation. Commonwealth Bureau of soils, Technical communication 51. Harpenden.

PLOEY, J. DE, 1965. Position géomorphologique, genèse et chronologie de certains dépôts superficiels au Congo occidental. Quaternaria 7: 131-154.

RAWITSCHER, F. 1946. Die Erschöpfung tropischer Böden infolge Entwaldung. Acta trop. 3: 211-241.

RICHARDS, P.W. 1936. Ecological observations on the rain forest of Mount Dulit, Sarawak. J. Ecol. 24: 1-37; 340-360.

RICHARDS, P.W. 1952. The Tropical Rain Forest. Cambridge.

RICHTER, J. 1972. Ökologie und System. In: Belastung und Belastbarkeit von Ökosystemen: 41-48. Giessen.

RIEDL, R. 1972. Generelle Eigenschaften der Biosphäre. In: Belastung und Belastbarkeit von Ökosystemen: 9-17. Giessen.

ROBYNS, W. 1946. Statistiques de nos connaissances sur les spermatophytes du Congo Belge et du Ruanda-Urundi. Bull. Jard. bot. Brux. 18: 133-144.

ROHDENBURG, H. 1970. Morphodynamische Aktivitäts- und Stabilitätszeiten statt Pluvial- und Interpluvialzeiten. Eiszeitalter u. Gegenwart 21: 81-96.

SCHMITHÜSEN, J. 1968. Towards an ecological characterisation of the South American continent. In: Biogeography and Ecology in South America 18: 113-136.

SHANNON, C.E. & W. WEAVER, 1949. The Mathematical Theory of Communication. Urbana.

SIOLI, H. 1956. As águas da regiao do Alto Rio Negro. Bol. téc. Inst. Agron. Norte 32: 117-155.

SIOLI, H. 1964. General features of the limnology of Amazonia. Verh. intern. Ver. Limnol. 15: 1053-1058.

SIOLI, H. 1968a. Hydrochemistry and geology in the Brazilian Amazon region Amazoniana I: 267-277.

SIOLI, H. 1968b. Zur Ökologie des Amazonas-Gebietes. In: E.J. FITTKAU, J. ILLIES, H. KLINGE, G.H. SCHWABE, H. SIOLI (ed.): Biogeography and ecology in South America. Vol. 1: 137-170. Junk, The Hague.

STEENIS, C.G.G.J. VAN 1938. Recent progress and prospects in the study of the Malaysian flora. Chron. Bot. 4: 392-397.

STEGMÜLLER, W. 1970. Einige Beiträge zum Problem der Teleologie und der Analyse von Systemen mit zielgerichteter Organisation. In: Aufsätze zur Wissenschaftstheorie: 21-56. Darmstadt.

THIENEMANN, A. 1918. Lebensgemeinschaft und Lebensraum. Naturw. Wochenschr. N.F. 17: 282-290, 297-303.

VOLTERRA, V. 1926. Variazioni e fluttuazioni del numero d'individui in specie animali conviventi. Mem. Accad. Lincei 2: 31-113.

WALTER, H. 1936. Nährstoffgehalt des Bodens und natürliche Waldbestände. Forstl. Wschr. Silva 24: 201-205; 209-213.

WIENER, N. 1948. Cybernetics. New York.

ZINDEREN BAKKER E.M. van 1967. Palynology and stratigraphy in sub-Saharan Africa. In: W.W. BISHOP & J.D. CLARK (ed.), Background to Evolution in Africa: 371-374. Chicago.

Anschrift des Verfassers: Prof. O. FRÄNZLE, Geographisches Institut der Universität Kiel, 23 Kiel 1, Olshausenstraße 40/60.

LA BIOGÉOGRAPHIE DES AMPHIBIENS DE GUYANE FRANÇAISE

par J. LESCURE

Abstract

The climatic zones and the ecological milieux of French Guiana are described. The geographic distribution of Amphibians of the marshes and savanahs of the coastal plain and the Rain forest is studied. Some forest species (*Dendrobates tinctorius, Allophryne ruthveni, Atelopus flavescens, Hyla proboscidea, H. ornatissima*) are exclusively guianan, others are common to Guianas and Amazon. Some Amazonian species (*Dendrobates quinquevittatus, Hyla reticulata, H. brevifrons*) are known only in the South-East and East of French Guiana. A speciation centre must have existed in Guianas, doubtless in a forest refuge (sensu HÄFFER) during arid phases. The French Guiana is a region of transition, the Amazonian influence that is preponderant in the South and East, diminishes gradually towards the North and West of the country.

Introduction

Au point de vue faunistique, la Guyane française a été beaucoup moins prospectée que les autres Guyanes bien que sa situation géographique à l'est du plateau guyanais et à proximité de l'Amazone est d'un grand intérêt pour la biogéographie de l'Amérique du Sud. Celle du complexe amazonien vient d'être renouvelé dans ses conceptions. En effet pour expliquer la variété et la richesse en Oiseaux de L'Amazonie, HÄFFER (1969) a émis l'hypothèse de refuges forestiers devenus des centres de spéciation durant des phases de la période postglaciaire, un des centres était situé dans la région guyanaise. Par d'autres méthodes, (MÜLLER (1969), VANZOLINI & WILLIAMS (1970) ont abouti aux mêmes suppositions.

A part un bref commentaire de PARKER (1935), la biogégraphie des Amphibiens des Guyanes n'a pas encore été étudiée. Or grâce à un abondant matériel que nous avons récolté depuis plusieurs années pour un inventaire des Batraciens de la Guyane française, les aires de répartition des espèces présentes en ce pays peuvent être établies.

Nous nous proposons d'expliquer l'existence d'espèces exclusivement guyanaises par l'hypothèse des refuges forestiers de HÄFFER, l'emplacement qu'aurait occupé ce refuge en Guyane française peut être localisé. Il est probable aussi que les différences de climat constatées en Guyane entrainent l'inégalité de la répartition des Amphibiens de L'Amazonie en cette région.

Régions climatiques et zones biogéographiques

Le climat guyanais

La Gyane française est située géographiquement mais non écologiquement en pleine zone équatoriale entre le deuxième et le sixième parallèle Nord. Elle est alternativement sous l'influence des alizés du nord-est et des alizés du sud-est. Les alizés se rencontrent dans la zone intertropicale de convergence (ZIC) qui est une zone de basses pressions relatives se déplaçant entre le troisième parallèle Sud et le quinzième parallèle Nord. La ZIC passe sur la Guyane en décembre et janvier puis remonte et stationne au-dessus du pays en mai et juin, mois où l'on enregistre les maxima de pluies.

Ces facteurs déterminent l'apparition de deux saisons principales: une saison des pluies s'étendant généralement du 15 novembre au 15 août et où domine l'alizé du nord-est et une saison sèche du 15 août au 15 novembre où domine l'alizé du sud-est.

Les pluies sont importantes, violentes et irrégulières aussi bien dans leur total annuel que dans leur répartition mensuelle géographique (moyenne annuelle sur 30 ans à Cayenne-Rochambeau; 3.750mm. ± 1250mm.).

La ZIC atteint sa position la plus méridionale en février-mars. Si à ce moment-là, l'alizé du nord-est n'est que modérement humide et instable, les journées sont ensoleillées d'où l'appellation de 'petit été de mars' employée en Guyane. Si cette période se prolonge comme en 1973, le déficit en pluie déclenche aussitôt un retour à la sécheresse particulièrement en savane.

Les pluies viennent le plus souvent de la mer, elles s'atténuent d'est en ouest et gagnent l'intérieur du pays en remontant le cours des fleuves qui sont dans l'axe des vents (Oyapok, Approuague, Comté, Maroni). Elles sont retardées dans leur progression par des massifs côtiers qui ne dépassent pas 300m et par d'autres massifs un peu plus élevés situés dans le centre nord du pays. Une autre onde de pluie moins importante vient du sud du pays et s'arrête vers le nord contre des montagnes du centre sud du pays (Mts du Camopi, Mts de Saül, Mts Attachi-Bacca, 700 à 900m.). La sécheresse emprunte ces mêmes itinéraires pour pénétrer et se développer en Guyane.

La saison sèche n'est pas caractérisée par une absence de pluie mais par une faible pluviométrie généralement en fin d'après-midi. Seules certaines régions peuvent ne pas recevoir d'eau pendant plus d'un mois. Pour mieux délimiter les périodes sèches, HOOK (1971) a utilisé dans ses travaux le concept de 'mois écologiquement sec'; suivant la convention généralement admise, il a considéré comme écologiquement sec un mois pendant lequel il tombe moins de 5cm. d'eau.

Régions climatiques

Au point de vue climatique la Guyane française est une zone de transition entre le Brésil côtier et les autres Guyanes. Si on examine la pluviométie des régions avoisinantes, on constate trois tendances: a. une pluviométrie abondante du Brésil côtier à l'est des Guyanes avec une moyenne annuelle de 3.500mm. environ.
b. une faiblesse relative des précipitations du Surinam central (moyenne à Paramaribo: 2.157mm).
c. des précipitations sensiblement réduites au sud du Tumuc-Humac, qui font écran aux alizés: 1500mm environ dans cette partie du Brésil.
Ces tendances se retrouvent en Guyane française et permettent de délimiter des régions climatiques.

I. Région côtiere

A une *bande côtière* (Ia) dont la profondeur est de l'ordre de 15 km on peut ajouter une '*zone médiane*' (Ib) large d'une centaine de kilomètres. On peut alors découper cette région en quatre groupes climatiques.

1. Le *groupe 'est'* caractérisé par une pluviométrie forte: moyenne annuelle de 3633mm., 221 jours de pluie et 1,7 mois écologiquement sec.

2. Le *groupe de Roura* qui correspond aux montagnes côtières, est l'endroit le plus arrosé de toute la Guyane française: 3959mm/an, 253 jours de pluie et 0.7 mois écologiquement sec. Les pluies peuvent atteindre 8000mm/an au sommet de certaines montagnes.

3. Le *groupe de Cayenne* qui comprend encore quelques montagnes côtières fait la transition entre le groupe de Roura et celui de l'ouest: 3187mm/an de pluie, 227 jours de pluie, 1,3 mois écologiquement sec.

4. Le *groupe 'ouest'* dont la partie côtière englobe toute la zone des savanes: 194 jours de pluie, 2615 mm/an et 2,1 mois écologiquement secs. A l'intérieur, le climat est le plus régulier de la Guyane: 2500 à 3500 mm an de pluie et 0,6 mois écologiquement sec entre la Mana et la Maroni.

II. Région de L'Intérieur

IIa. Zone Sud.

Elle fait suite à la zone médiane. Son climat varie d'est en ouest. Dans le bassin de l'Oyapok et de l'Approuague les pluies sont abondantes et on remarque souvent l'absence du petit été de mars, cependant la saison sèche est plus marquée (Camopi: 230 jours de pluie, 2716mm/an et 1 mois sec). Au centre de la Guyane, la région de Saül est après le Tumuc-Humac le principal château d'eau du pays (235 jours de pluie, 2400mm/an et 0,6 mois sec). Dans la région du Haut-Maroni la saison sèche est plus marquée qu'à l'est.

IIb. Zone de l'extrême-sud

La pluviométrie y est la plus faible de toute la Guyane (moins de 2000mm/an), la saison sèche y est assez forte (1,44 mois sec). C'est exclusivement dans cette zone que l'on trouve des affleurements dénudés de Granite isolés en pleine forêt dense. Actuellement la végétation forestière semble reconquérir petit à petit les pentes de ces Inselbergs, elle est encore très clairsemée et y évolue en fourrés et en forêt plus sèche et plus claire que la forêt dense environnante.

Milieux écologiques

La *bande côtière* (Ia) comprend une plaine côtière récente et une plaine côtière ancienne. La première est constituée d'alluvions marines argileuses. Elles s'étendent de l'embouchure de l'Amazone à celle de l'Orénoque sur près de 2000 km de long sans autre discontinuité que celle créée par 'l'île' de Cayenne, le seul endroit où le socle précambrien affleure jusqu'à la mer. La largeur de cette plaine est très variable, restreinte à quelques kilomètres entre Cayenne et Organabo, elle s'élargit vers Mana et surtout à l'est de Cayenne où elle atteint 50 km à la pointe Béhague.

Couverte principalement de marécages, cette plaine a une superficie de 370.000 ha en Guyane française, 1550.000 ha au Surinam et 1200.000 ha en Guiana (cf. Fig. 1).

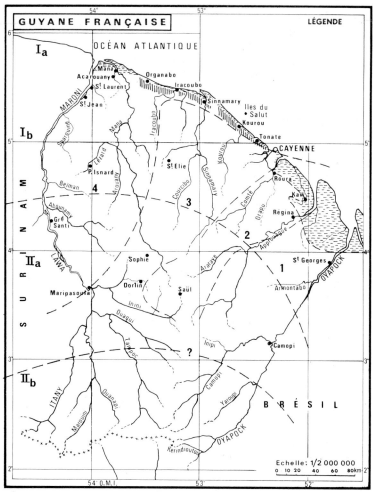

Fig. 1. Régions climatiques et zones biogéographiques de la Guyane française. Ia. Bande côtière. Ib. Zone médiane: 1. groupe 'est'; 2. groupe de Roura; 3. groupe de Cayenne; 4. groupe 'ouest'. IIa. Zone sud. IIb. Zone de l'extrême-sud. Traits horizontaux: Marécages côtiers. Traits verticaux: Savanes.

Un fait très remarquable de ces marécages côtiers est le groupement d'espèces végétales en associations caractéristiques, il est alors facile d'esquisser un certain nombre de paysages végétaux:

— La mangrove à Palétuviers soit en bord de mer avec *Avicennia nitida* soit en estuaire avec *Rhizopora*;
— La mangrove décadente;
— les palmeraies marécageuses ou Pinotières à *Euterpe oleracea*;

— Les prairies marécageuses à *Echinochloa polystachia* (graminée) et celles à *Eleocharis* ou à *Heliconia psittacorum* nommées localement 'pripris'.

En arrière de la plaine côtière récente, la *plaine côtière ancienne* (old coastal plain des Guyanes) se rencontre avant d'atteindre le socle précambrien. Elle est formée d'alluvions marines anciennes souvent très sableuses, c'est le domaine des savanes. En Guyane française celles-ci sont uniquement localisées dans la partie centrale de la bande côtière entre Cayenne et Organabo (cf. fig. 1). Elles s'étendent sur 150 km avec une largeur variant de quelques centaines de mètres à 15 km sur une superfice de 75.000 ha. Elles sont cloisonnées par des galeries forestières et parsemées d'îlots forestiers; transversalement elles sont morcelées par un grand marais (marais Yi-Yi) et par les estuaires des fleures. HOOK (1971) a distingué deux types de savanes comprenant chacun un certain nombre de groupements végétaux: La savane haute dont la végétation ne dépasse pas 1,50 m est soit arbustive soit herbeuse; la savane basse à graminées et à cypéracées atteignant 30 à 60cm est herbacée, arbustive, marécageuse ou à nanophanérophytes. La forêt dense équatoriale fait suite aux savanes et aux marécages côtiers et recouvre la plus grande partie de la Guyane française. Cette forêt est primaire excepté dans la région de Cayenne et autour des endroits habités. Les mares, les petits marécages, les rives des fleuves et des rivières constituent des biotopes particuliers dans ce milieu apparemment homogène. Au-dessus de 600m, la végétation acquiert une physionomie annonçant la forêt d'altitude. Une zone située au nord d'une ligne Iracoubo-Maripasoula a une flore sembable à celle du Surinam (cf. fig. 1).

Amphibiens et Zones biogéographiques

Amphibiens des marécages côtiers

Au point de vue biogéographique on peut distinguer deux zones dans les marécages côtiers de Guyane française. L'une à l'est de Cayenne a des paysages caractéristiques et une faune à affinité amazonienne, c'est la région des grands marais de Kaw. L'autre, à l'ouest de Cayenne a une flore et une faune très proches de celles du Surinam.

Les grand marais de Kaw qui se prolongent jusqu'à l'estuaire de l'Oyapok, sont en continuité avec l'estuaire de l'Amazone par les marais de Ouassa, du Cachipour, de Carsevenne et d'Amapa (Territoire de l'Amapa). C'est le domaine du grand Caïman *Melanosuchus niger* et de la Tortue Mata-mata *Chelys fimbriata*, on y voit parfois des Lamentins. Cette faune ne remonte pas l'Oyapok mais passe directement de la rivière Ouassa a l'Approud gue par les Marais.

Un des paysages typiques de cette région est la prairie marécageuse à *Echinochloa polystachia*, Graminée fourragère, associée à un *Cyperus*, elle est appelée localement 'savane inondée'. Les 'savanes' Gabrielle, Angélique et de la vallée de Kaw sont ainsi noyées sous un à deux mètres d'eau en saison des pluies alors qu'en saison sèche on peut marcher —a travers des herbes de plus de deux mètres. Dans la savane Gabrielle, j'ai trouvé *Hydrolaetare schmidti*, Leptodactylidé aquatique, connu seulement de Létitia et du Matto-Grosso. Dans les herbes, à quelques centimètres au-dessus de l'eau, vit *Hyla boesemani*, espèce récoltée entre Bélem et le Surinam.

Dans les buissons de Mélastomacées, sur la bordure de ces 'savanes inondées', on entend dès la tombée de la nuit: *Hyla rostrata*, espèce particulière des milieux ouverts de Panama, du Vénézuela et de Colombie, on y entend aussi *Hyla egleri*.

Dans 'l'île' de Cayenne et au nord-ouest de cette ville les marécages côtiers sont de moins grande dimension. On distingue des 'pripris a joncs' genre de prairies marécageuses à *Eleocharis* dominant. C'est le biotope préféré d'une petite *Hyla* du groupe *rhodopepla* que j'ai seulement trouvée aux environs de Cayenne. Contrairement aus 'savanes inondées', l'eau ne recouvre jamais totalement la végétation de ces marais. Le pripri à *Heliconia psittacorum* dominant est encore moins profond, j'y ai capturé plusieurs espèces communes à l'Amérique tropicale: *Hyla punctata H. rubra, Pipa pipa* mais aussi *Hyla egleri, Pseudis paradoxa paradoxa* sous-espèce particulière aux Guyanes (GALLARDO, 1961) et *Hyla raniceps* qui n'existe pas au Surinam et ne remonterait pas au delà de Kourou.

Biogéographiquement les marais de 'l'Anse' de Sinnamary présentent quelques particularités amazoniennes, on y a récolté la Tortue Mata-mata dont c'est à notre avis la station la plus septentrionale dans les Guyanes.

Amphibiens des savanes

Les espèces rencontrées dans les savanes basses herbacées sont les mêmes que dans les autres savanes de l'Amérique du Sud tropicale: *Leptodactylus ocellatus, Bufo marinus* et *Leptodactylus fuscus* très répandu dans les 'savanes à touradons'. Dans les îlots forestiers à roche affleurante gîte *Bufo granulosus merianae*, sous-espèce particulière aux Guyanes (GALLARDO, 1965).

Dans les savanes basses marécageuses et les buissons près des ruisseaux, on retrouve la faune des marécages. Cependant à Iracoubo au nord-ouest de Cayenne, je n'ai pas récolté *Hyla gr. rhodopepla* mais Hyla *microcephala mesera* commune à Paramaribo. A Kourou, j'ai encore trouvé *Hyla raniceps* mais je n'ai pas vu *Hyla crepitans*.

Amphibiens de la forêt

La biogéographie des espèces forestières est plus complexe que celle des espèces des savanes ou des marécages. Dans ce biotope apparemment uniforme, la faune en Amphibiens est très diversifiée. Cependant grâce au matériel abondant que j'ai récolté récemment en Guyane française et aux travaux de HOOGMOED (1969, a et b) certains faits être dégagés.

Parmi les 65 espèces inventoriées jusqu'à maintenant dans la forêt de Guyane française, 49 vivent aussi en Amazonie et 16 sont guyanaises. Une étude d'ensemble de leur répartition géographique serait prématurée actuellement car la révision systématique de certains de leurs groupes n'a pas encore été effectuée. Nous nous bornerons à exposer plusieurs exemples caractéristiques.

Les Dendrobates. *Dendrobates tinctorius* est un Amphibien terrestre, diurne et aux couleurs brillantes avertisseuses que l'on trouve assez facilement sur le sol de la forêt. C'est une espèce uniquement guyanaise qui vit au sud et à l'ouest de la Guiana, au Surinam, en Guyane française et en Amapa. Elle déborde peut-être le Tumuc-Humac puisqu'elle a été trouvée par Aubert de la Rüe entre Caïman et Ourouareu, aux sources de l'Oyapok. Elle est abondante en altitude (500-700m) et colonise les sommets des petites montagnes où la température est plus fraîche et la forêt plus humide (Mts Attachi-Bacca, Mts Cacao, Mts Galbao en Guyane française).

Des variations géographiques peuvent être décelées dans cette espèce. Sur la

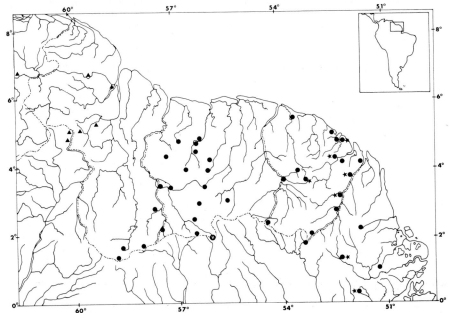

Fig. 2. Répartition du genre *Dendrobates* dans les Guyanes *D. tinctorius* (cercles noirs). *D leucomelas* (triangles). *D azureus* (étoile blanche dans un cercle). *D. quinquevittatus* (étoiles noires).

côte dans la région de Cayenne et de Roura, je n'ai récolté que des spécimens de petite taille, à deux bandes fines latérodorsales et une ou deux bandes transversales blanches. La couleur de fond est bleu-noir, le ventre et la gorge sont parcourus de fines bandes bleues très anastomosées entre elles, le dessus des membres est bleu.

Vers le nord en Guiana, les dessins bleus ventraux passent sur le dos, il n'y a plus de bandes dorsales blanches ou jaunes. Le cas extrême de cette évolution est celui de *Dendrobates azureus* HOOGMOED, 1969, qui est entièrement bleu. Cette espèce endémique est isolée géographiquement car elle vit dans les bosquets à l'intérieur de la savane de Sipaliwini, au sud du Surinam.

A l'intérieur des Guyanes, particulièrement en altitude, les bandes dorsales de *D. tinctorius* s'élargissent et deviennent des taches recouvrant la tête et se terminant en V dans le dos, elles débordent parfois sur les flancs et le ventre, leur couleur est jaune. Les flancs, les pattes, le ventre et le bas du dos sont noirs. Dans la région du Haut-Oyapok (Trois-Sauts) les bandes ou les taches dorsales sont oranges.

Au nord-ouest de la Guiana (bassin des rivières Cuyuni et Mazaruni) on ne trouve plus *D. tinctorius* mais *Dendrobates leucomelas* qui habite la 'Guyane' vénézuélienne, le nord-est de la Colombie et le nord du Brésil. Au sud, dans la région de Para, *Dendrobates galactonotus* fait suite à *D. tinctorius*, il lui est très proche morphologiquement.

Dendrobates quinquevittatus récolté surtout en Equateur et au Pérou mais aussi à Manaus, Benjamin Constant, Rio Madeira et Bélem n'a encore jamais été signalé en Guiana et au Surinam. Je ne l'ai pas trouvé à l'ouest de la Guyane française mais dans l'est et au centre (Saül, Moyen-Oyapok, Approuague, Montagne de Roura). Cette petite espèce amazonienne qui est rare en Guyane française et dans l'Amapa

91

semble donc avoir pénétré dans les Guyanes par le sud-est et n'aurait pas franchi le Tumuc-Humac.

Les aires de répartition des *Dendrobates* du nord-ouest de l'Amérique du Sud correspondent aux 'refuges' qui auraient existé au pleistocène et seraient devenus ensuite des centres de spéciation (HÄFFER 1969) *Dendrobates leucomelas* viendrait du refuge de l'Iméri, *D. tinctorius* de celui des Guyanes, *D.galactonotus* de celui de Para et *D. quinquevittatus* de celui du Napo (Equateur).

Les *Atelopus. Atelopus flavescens* DUMERIL et BIBRON, 1841, est une espèce exclusivement guyanaise, elle n'a été récoltée jusqu'à présent qu'en Guyane française et en Amapa. Pour éviter toute confusion, il faut souligner qu'*Atelopus flavescens* BOULENGER, 1881, est très différent s'*Atelopus flavescens* DUMÉRIL et BIBRON, c'est une sous-espèce d'*Atelopus pulcher* BOULENGER: *Atelopus p. hoogmoedi* (LESCURE, 1972, 1973).

Atelopus flavescens DUM. et BIBR. ne se trouve pas dans l'intérieur de la Guyane française, il est cantonné à la zone médiane particulièrement à la région de Rochambeau, Roura, Kaw et l'Approuague (région Ib, groupe de Roura). Geay l'avait récolté à la rivière Lunier dans l'Amapa. *At. flavescens* est différent de toutes les autres espèces excepté *At. pulcher* qui a aussi un tympan interne, une oreille moyenne et le premier orteil inclus dans la peau.

Atelopus franciscus nov. sp., plus petite et de couleur différente est localisé au bassin du Sinnamary.

L'intérieur des Guyanes est occupé par *Atelopus pulcher hoogmoedi*, sous-espèce dérivée d'*Atelopus pulcher* BOULENGER de L'Equateur et de l'est du Pérou. Une forme caractéristique avec ses larges bandes dorsales noires et sa couleur jaune vif est particulière au bassin du Haut-Maroni. *Dendrophryniscus minutus* récolté récemment dans le Haut-Maroni et dans le Haut-Oyapok est une espèce du complexe amazonien. *Dendrophryniscus proboscideus* est connu de Guiana.

Les Hylidés. *Allophryne ruthveni* est une petite Rainette exclusivement guyanaise. GAIGE (1926) a décrit ce genre monotypique à partir de spécimens provenant de l'ex-Guyane britannique. BOKERMAN (1958) l'a trouvé à Serra do Navio (Amapa) et HOOGMOED (1969a) en a récolté récemment au Surinam. Je l'ai capturé en Guyane française où il n'avait pas encore été signalé et j'ai pu ainsi compléter la carte de sa répartition établie par HOOGMOED. A part quelques prises dans la région côtière d'Albina (Surinam), *Allophyrne ruthveni* a été découvert dans l'intérieur de Guyanes. Cette espèce a été classée dans les Hylidés par LYNCH FREEMAN (1966) mais sa place indécise dans la classification peut signifier que elle est dérivée d'une souche guyanaise assez ancienne.

Hyla ornatissima NOBLE est une espèce très rare. RIVERO (1961) pense même qu'elle serait une variété d'*Hyla granosa*, espèce commune à l'Amazonie et aux Guyanes. Il est vrai que des *Hyla granosa* de la collection de l'American Museum (AMNH 39985) ont quelques taches dorsales qui rappeleraient un peu les dessins caractéristiques de *H. ornatissima*. J'ai examiné les cinq spécimens connus de cette espèce guyanaise, leurs dessins dorsaux très particuliers sont exactement sembla-bles malgré les distances qui séparent les stations. D'après NOBLE (1923) sa coloration ornementale in vivo parait aussi très différente de celle d'*Hyla granosa*. Jusqu'à présent on n'a trouvé *H.ornatissima* qu'à Meanu (Mazaruni Riv., Guiana), vers les sources du Gran Rio (Surinam), à Yanioué (Haut-Camopi, Guyane française), au Rio Camaipi et au Rio Amapari dans l'Amapa. (cf. fig. 4).

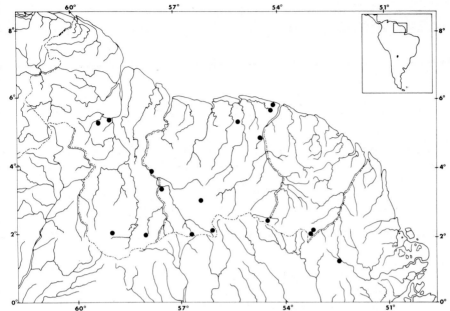

Fig. 3. Répartition de *Alloprhryne ruthveni* (cercles) dans les Guyanes.

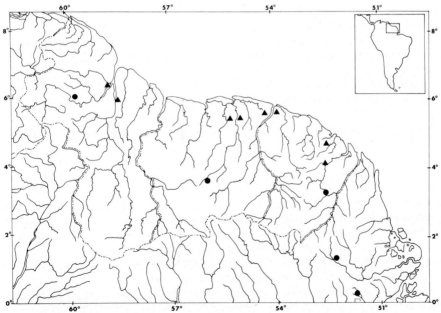

Fig. 4. Répartion de *Hyla ornatissima* (cercles) et *Hyla granosa* (triangles) dans les Guyanes.

93

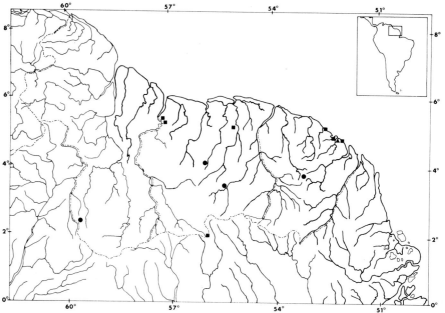

Fig. 5. Répartition des *Hyla* du groupe *rostrata* dans les Guyanes *H. proboscidea* (cercles). *H. Rostrata* (triangles). *H. egleri* (carrés).

Hyla proboscidea est une autre Rainette très rare et très singulière avec son rostre cutané. BRONGERSMA (1933) l'avait décrite à partir d'un exemplaire de Gran Rio (Surinam), quatre spécimens proviennent de Shudikar-Wau (Guiana) et j'en ai capturé un à Saül (Guyane française). *H. proboscidea* est une *Hyla* du groupe *rostrata* dont DUELLMAN (1972) vient de faire la révision. Le manque de corrélations des différences interspécifiques des six espèces de ce groupe suggère que leur spéciation s'est effectuée dans des régions isolées les unes des autres. Or cinq d'entre elles ont un habitat correspondant aux refuges supposés par HÄFFER (1969). *H.proboscidea*, comme L'indique sa répartition décrite ci-dessus, proviendrait du refuge guyanais, *H.egleri* de celui de Bélem, *H.epacrorphina* de celui de l'est du Pérou, *H. garbei* de celui du Napo, *H. boulengeri* de celui du Choco et peut-être aussi de celui du Costa-Rica. *Hyla rostrata* qui semble l'espèce la plus primitive du groupe n'est pas forestière mais habite des milieux ouverts humides du nord de l'Amérique du Sud et de l'Amérique centrale. Je l'ai trouvée en Guyane française au bord des savanes inondées de la Crique Gabrielle. *Hyla egleri* est commune dans les buissons des marécages côtiers; à la Crique Gabrielle, elle est dans le même biotope que *H. rostrata* et peut donc être en concurrence écologique avec elle.

Hyla boesemani GOIN est une petite Rainette commune dans les marécages côtiers. Je l'ai trouvée aussi dans des petits marécages de forêt à Trois-Sauts (Haut-Oyapok) et dans des petites savanes artificielles à Saül. C'est une espèce récoltée jusqu'à présent au Surinam, en Guyane française et à Bélem et qui ne semble pas très liée aux autres *Hyla* du même groupe *rubra*.

Les Microhylidé. Malgré l'absence de travaux de synthèse sur cette famille pour l'Amérique du Sud et la rareté des spécimens en collection, on peut affirmer

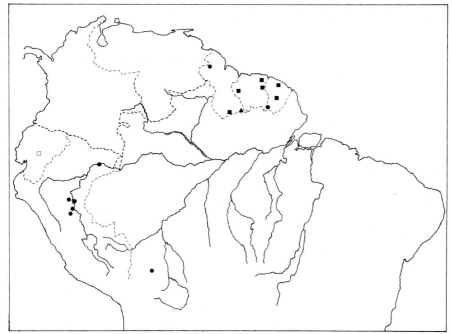

Fig. 6. Répartition de *Hamptophryne boliviana* (cercles) et de certains *Chiasmocleis* en Amérique du Sud. *Ch. shudikarensis* (carrés noirs). *Ch. ventrimaculata* (carré blanc). *Ch. hudsoni* (triangle).

qu'*Hamptophryne boliviana* est une espèce du complexe amazonien présente en Guiana, Surinam, Guyane française, Haute-Amazonie et au sud-ouest du Brésil. *Chiasmocleis ventrimaculata* ANDERSON de l'Equateur soit synonyme de *Ch. shuddikarensis*, alors celui-ci récolté jusqu'à présent dans les trois Guyanes serait amazonien. *Synapturanus microps* (DUMERIL et BIBRON) est une espèce guyanaise, mais le nom est impropre. Le type de *Synapturanus microps* est de Rio de Janeiro et représente ce qu'on nomme actuellement *Myersiella subnigra*.

Ctenophryne geayi, espèce de l'Amazonie, de l'est de la Colombie et du Vénézuela a été recuelli en Guiana et au Surinam mais pas en Guyane française. Vénézuela a été receuilli en Guiana et au Surinam mais pas en Guyane française. *Elachistocleis ovalis* est commun à l'Amérique du Sud tropicale mais il y a sans doute plusieurs espèces sous ce nom (NELSON, commun. pers.). *Otophryne robusta* que l'on disait endémique du Mont Roraima est une espèce commune dans la forêt guyanaise mais difficile à capturer. J'en au trouvé une dizaine dans le Haut-Oyapok et je l'ai entendue aux Mts Attachi-Bacca et aux Mts Galbao, elle a été signalée en Amapa et en Amazonie colombienne. Une sous-espèce d'altitude découverte au Chemanta-tepui au Vénézuela a été décrite par RIVERO (1967). Les espèces de l'Amazonie. Dans les exemples décrits ci-dessus, nous avons déjà cité des espèces vivant en Amazonie et en Guyane. Quelques unes ont une vaste distribution géographique, elles débordent l'Amazonie et envahissent d'autres forêts tropicales humides de l'Amérique du Sud et parfois de l'Amérique centrale. C'est le cas notamment de *Leptodactylus pentadactylus*,

95

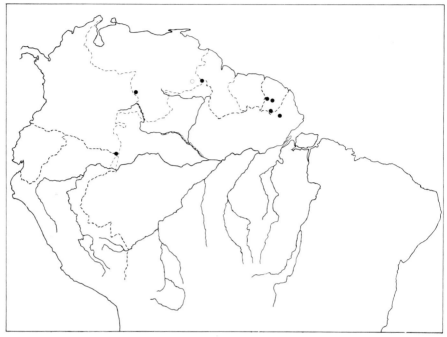

Fig. 7. Répartition de *Otophryne robusta* en Amérique du Sud. *O. r. robusta* (cercles). *O. r. steyermarki* (cercle blanc).

Phrynohias venulosa, Hyla boans, Hyla rubra et *Centronella fleischmani* (les trois dernières n'atteignent pas le sud du Brésil). *Bufo marinus,* remplacé dans le sud par *Bufo paracnemis,* remonte au nord jusqu'en Floride, mais ce grand Crapaud n'est pas une espèce de la forêt primaire, il est ubiquiste et vit dans les lieux habités, les plantations, la végétation secondaire, les savanes plus ou moins marćageuses et le long des fleuves.

Certaines espèces de l'Amazonie comme *Hyla reticulata, Physaelemus petersi, Dendrobates quinquevittatus* sont apparemment absentes de la région du Maroni, du Surinam et de Guiana. En Guyane française, je ne les ai trouvées que dans les régions de l'Oyapok, de Saül et de l'Approuague. Ces espèces ne sont donc pas guyanaises. Provenant probablement du refuge du Napo, elles ont sans doute été gênées dans leur extension vers les Guyanes par un corridor situé derrière le Tumuc-Humac où la pluviométrie est plus faible et où la forêt n'a pas encore retrouvé toute se stabilité (MÜLLER & SCHMITHUSEN, 1970). Elles sont donc arrivées dans les Guyanes françaises et brésiliennes (Amapa) par le sud-est où les pluies sont plus fortes.

Un cas extrême est celui de *Rana palmipes* qui est commune dans les forêts tropicales du nord de l'Amérique du Sud et a aussi pénétré en Amérique centrale. Elle existe en Guiana et à Boa Vista (Roraima), elle est abondante sur les rives de L'Oyapok, de la Comté et du Sinnamary mais elle n'a pas été capturée vers le Maroni et au Surinam. Cependant un facteur écologique a dû jouer un rôle, cette grenouille semi-aquatique gitant au bord des fleuves et des rivières, la ligne de partage des eaux au Tumuc-Humac a peut-être été un obstacle dans sa progression.

96

L'existence de plusieurs centres d'origine pour les Amphibiens actuels de la Guyane française est démontrée également par la présence en forêt d'espèces sympatriques, bien isolées sexuellement puisque leurs chants sont très distincts. Elles paraissent de temps en temps en concurrence pour la prédation car elles sont de taille voisine et vivent parfois dans les mêmes niches écologiques. Le long des fleuves, *Hyla geographica, H. calcarata* et *H. gr. fasciata* logent aux mêmes endroits dans la végétation luxuriante qui borde les rives. Cependant *H. geographica* est uniquement forestière, tandis que *H. calcarata* et *H. gr. fasciata* envahissent les bois inondés de la côte mais y demeurent à des hauteurs différentes. Deux espèces de *Bufo* du groupe *typhonius* habitent la litière de la forêt mais une seule déborde dans les bois secondaires de la côte. Dans le Haut-Oyapok, *Hyla reticulata* et *Hyla leucophyllata* sont côte à côte dans les marécages de forêt. Tous ces Amphibiens ont dû accomplir leur spéciation durant les périodes d'isolement géographique (phases de climat sec) et venir en contact ou se superposer quand les conditions de climat humide se sont rétablies.

Le refuge guyanais. Peut-on alors délimiter dans la forêt de Guyane française la zone que le 'refuge' guyanais aurait occupé lors des phases de climat sec? Pour y aboutir, j'ai repris l'hypothèse de HÄFFER (1969) suivant laquelle les aires actuelles de pluviométrie maximale doivent correspondre à des zones qui étaient assez pluvieuses pour permettre la persistance de la persistance de la forêt. En effet, les faits orographiques qui provoquent en ce moment les inégalités dans la pluviométrie du complexe amazonien n'on pas dû changer sensiblement durant le pléistocène.

La partie du refuge guyanais en Guyane française doit se localiser dans les régions de la Comté (pluviométrie maximale), de Saül et dans la zone centrale de l'extrême-sud entre le Camopi et le Tampoc jusqu'au versant nord du Tumuc-Humac. (c.f. fig 1). Il est possible qu'elle soit rattachée à la partie surinamienne du refuge par les Mts Attachi-Bacca et les Mts Orange du Surinam. Les régions de la Comté et de Saül sont connues pour la richesse et la variété de leur flore et de leur faune, celle de l'extrême-sud n'a pas encore été prospectée. Ces trois secteurs sont montagneux et sont le point de départ de plusieurs fleuves et de nombreuses rivières.

Le refuge guyanais a été isolé géographiquement de l'Amazonie et est devenu ensuite un centre de spéciation d'une faune guyanaise. On peut raisonnablement affirmer qu'il l'a été pour des Amphibiens tels que *Dendrobates tinctorius, Atelopus flavescens, Atelopus franciscus, Allophryne ruthveni, Hyla proboscidea, Hyla ornatissima* et *Chiamocleis hudsoni.*

Conclusion

L'étude de la répartition géographique des Amphibiens récoltés en Guyane française nous a permis de dégager un certain nombre de faits.

Certaines espèces forestières sont endémiques, elles vivent généralement dans les régions de l'intérieur (région climatique II, en Guyane française). Quelques unes, des Hylidés, sont très rares.

Des espèces sont communes aux Guyanes et à tout le complexe amazonien: certaines sont répandues dans toutes les régions tropicales humides de l'Amérique du Sud et débordent en Amérique centrale; d'autres habitent les savanes, vivent aussi en Argentine et au sud du Brésil.

Des espèces forestières connues dans la Haute-Amazonie ont été récoltées dans les régions de l'intérieur des Guyanes mais pas dans la région côtière.

Des espèces forestières ou marécageuses capturées dans l'Amapa et en Guyane française principalement dans la région de l'Oyapok, de Saül et de l'Approuague sont apparemment absentes du Surinam et de Guiana.

Des espèces sympatriques, voisines morphologiquement, sont parfois en concurrence écologique. Elles doivent donc venir de centres de spéciation différents et se chevaucher dans leurs aires de répartition.

On peut donc affirmer qu'au point de vue faunistique, les Guyanes ne sont pas une simple partie de l'Amazonie mais une région particulière qui possède sa propre faune malgré l'absence de véritables barrières orographiques.

Cette faune s'est constituée quand la forêt guyanaise a été isolée de la forêt amazonienne et provient d'un centre de spéciation qui avait servi de refuge pendant les phases arides de la période postglaciaire. Suivant les statistiques de pluviométrie, la répartition de la flore et de la faune en Amphibiens, on peut supposer qu'en Guyane française ce refuge était situé au nord du Tumuc-Humac entre le Tampoc et le Camopi, autour de Saül et entre la Comté et l'Approuague.

Cependant dans la phase humide actuelle de nombreuses espèces amazoniennes ont envahi la forêt guyanaise mais ne l'occupent pas totalement. Ces influences amazoniennes sont plus fortes en Guyane française qu'au Surinam. Ce pays, situé au centre des Guyanes, n'a de contact avec l'Amazonie que par le sud où la présence de savanes comme à Sipaliwini fait suggérer l'existence d'un climat plus sec et sans doute défavorable à l'émigration de certains Amphibiens de l'Amazonie.

En Guyane française, ces espèces pénètrent plus efficacement par le sud-est qu'à travers le Tumuc-Humac. Le pays apparait comme une région de transition. L'influence amazonienne prépondérante au sud et à l'est diminue en allant vers le nord et l'ouest du territoire. A l'est de Cayenne les grands marais de Kaw sont en continuité avec l'estuaire de l'Amazone. A l'ouest de Cayenne et particulièrement à partir d'Iracoubo, la flore et la faune en Amphibiens de la forêt, des marécages et des savanes sont semblables à celles du Surinam. (cf. fig. 1).

BIBLIOGRAPHIE

BOULENGER, G.A., 1882. Catalogue of the *Batrachia Salientia S. Ecaudata* in the collection of the British Museum. 2e édit. Londres. 503 p.

BRONGERSMA, L.D., 1933. Ein neuer Laubfrosch aus Surinam. *Zool. Anz.*, 103 (9/10): 267-270.

DUELLMAN, W.E. 1970. Hylid frogs of Middle America. Monograph Mus. Nat. Hist. Univ. Kansas. 2, 753 p.

DUELLMAN, W.E., 1971. A taxonomic review of South America Hylid frogs, genus Phrynohyas. *Occ. Pap. Mus. Nat. Hist. Univ. Kansas*. 4: 1-21.

DUELLMAN, W.E., 1972. South American frogs of the *Hyla rostrata* group *(Amphibia, Anura, Hylidae). Zool. Mededelingen* 47 (14): 177-192.

DUMERIL, A.M.C. & BIBRON, 1872. Erpétologie générale. Paris, T. 8, 792 p.

DUNN, E.R. 1949. Notes on the South American frogs of the family *Microhylidae. Am. Mus. Novitates.* 1419: 1-21.

FOUGEROUZE, J., 1965. Le climat de la Guyane française. Type de temps, Saisons et régions climatique. Monographies de la Météorologie nationale. 38, 36 p.

GAIGE, H.T., 1926. A new frog from British Guiana. *Occ. pap. Mus. Zool. Univ. Michigan* 176: 1-3.

GALLARDO, J.M., 1961. On the species of *Pseudidae (Amphibia, Anura). Bull. Mus. Comp. Zool.* 125 (4): 3-134.

GALLARDO, J.M., 1965. The species *Bufo granulosus* Spix (*Salientia, Bufonidae*) and its geographic variation. *Bull. Mus. comp. Zool.* 134 (4): 107-138.

HÄFFER, J., 1969. Speciation in Amazonian forest Birds. *Science* 165 (388): 131-137.

HOOGMOED, M.S., 1969. Notes on the Herpetofauna of Surinam II. On the occurence of *Allophryne ruthveni* Gaige (*Amphibia, Salientia, Hylidae*) in Surinam. *Zool. Med.* 44 (5); 75-81.

HOOGMOED, M.S., 1969. Notes on the Herpetofauna of Surinam III. A new species of *Dendrobates, (amphibia, Salientia, Dendrobatidae)* from Surinam. *Zool. Med.* 44 (9): 133-141.

HOOK, J., 1971. Les savanes guyanaises: Kourou. Essai de phytoécologie numérique. M⸍emoires ORSTOM 44, 251 p.

LESCURE, J., 1972. Contribution à l'étude des Amphibiens de Guyana française. I. - Notes sur *Atelopus flavescens* Duméril et Bibron et description d'une espèce nouvelle. *Vie et Milieu.* 23 (1-C): 125-141.

LESCURE, J., 1973. Présence d'une sous-espèce d'*Atelopus pulcher* (Amphibien, Anoure) dans les Guyanes: *Atelopus pulcher hoogmodi*. Bull. Mus. Hist. Nat. Paris, Zool. 108, no. 144.

LYNCH, J.D. & H.L. FREEMAN, 1966. Systematic status of a South American frog *Allophryne ruthveni* Gaige. *Univ. Kansas Publ. Mus. Nat. Hist.* 17 (10): 493-502.

MARIUS, Cl. & J.F. TURENNE, 1968. Problèmes de classification et de caractérisation des sols formés sur alluvions marines récentes dans le Guyanes. *Cah. ORSTOM sér. Pédol..* 6 (2): 151-201.

MÜLLER, P., 1969. Vertebratenfaunen brasilianischer Inseln als Indikatoren für glaziale und postglaziale Vegetationfluktuation. *Zool. Anz. Suppl. B.* 33, *Verh. Zool. Ges.* p. 97-107.

MÜLLER P. & J. SCHMITHUSEN, 1970. Probleme der Genese Südamerikanischer Biota. Fetschr. Gentz. Verl. F. Hirt, Kiel, p. 109-122.

NELSON, C.E., 1971. A Brazilian record for the Microhylid *Otophryne robusta*. *Herpetpologica* 27 (3): 234-325.

NOBLE, G.K., 1923. New Batrachians from the tropical research station Britsh Guiana. *Zoologica* 3 (14): 289-305.

PARKER, H.W., 1935. The frogs, lizards and snakes of British Huiana. *Proc. Zool. Soc. London* p. 505-530.

RIVERO, J.A., 1961. Salientia of Venezuela.*Bull. Mus. Comp. Zool.* 126 (I) 207 p.

RIVERO, J.A., 1967. A new race of *Otophryne robusta* Boulenger (*Amphibia, Salientia*) from the Chemanta-Tepui of Venezuela. *Carib. J. Sci.* 7 (3/4): 155-157.

TRUEB L. & W.E. DUELLMAN, 1971. A synopsis of Neotropical frogs genus *Osteocephalus.Occ. pap. Mus. Nat. Hist. Univ. Kansas* I: 1-47.

VANZOLINI P. & E.E. WILLIAMS, 1970. South American Anoles: The geographic differentiation and evolution of the *Anolis chrysolepsos* species group (*Sauria, Iguanidae*). *Arq. Zool.* 19: 1-298.

Anschrift des Verfassers:
Dr. L. LESCURE, Muséum national d'Histoire naturelle, Paris. France.

POSSIBILITIES AND PROSPECTS OF DRY-FARMING IN THE REGIONS OF NORTH-EAST BRAZIL WITH AN UNRELIABLE RAINFALL

F. CHRISTIANSEN-WENIGER[1]

Abstract

A study-tour to investigate possibilities of arable husbandry was made in 1972 through four states in the dry area of north-east Brazil. Although in these tropical areas the temperature conditions are relatively uniform the rainfall is spatially and temporally very uneven. Generally there is an alternation of rainy and dry seasons. Rainfall data for 40 stations shows the distributional patterns. Map 1 gives the average annual rainfalls. Map 2 gives the rainfall amounts in the dry year 1958. Map 3 shows for individual stations the differences between maximum and minimum annual rainfall expressed as a percentage of the mean. This shows the districts in which rainfall is particularly variable. Table 16 lists all places with under 500 mm average annual rainfall, and shows the average distribution of yearly falls per decade. It therefore also shows whether arable farming is economic. Finally it is shown which agricultural problems should be investigated by means of a uniform programme over a number of years in order that modern farming with the highest possible yields can develop.

In the States of Bahia, Ceara Maranhao, Paraiba, Pernambuco, Piaui, Rio Grande do Norte and Sergipe in north-east Brazil are extensive areas often subject to catastrophic drought. The resulting poor harvests and continuous shortages of water and fodder, which become accentuated during the dry period, have led to an urban migration of the rural population. In all, the nine states comprise an area of 1.5 million Km^2 with a population in 1971 of 28 millions.

Table 1. The states of north-east Brazil.

state	surface area Km^2	population	population per Km^2
Alagoas	27,731	1,606,165	57.9
Bahia	561,026	7,420,906	13.2
Ceara	148,016	4,440,286	30.0
Maranhao	328,663	2,883,211	8.8
Paraiba	56,372	2,383,518	42.3
Pernambuco	98,281	5,208,011	52.0
Piaui	250,934	1,735,568	6.8
Rio Grande do Norte	53,015	1,603,094	30.2
Sergipe	21,994	900,119	41.0
total	1,546,032	28,180,878*	18.3

*population figures relate to 1971.

1 I would like here to thank Der Deutschen Forschungsgemeinschaft, which made possible my journey to Brazil. Thanks are also due to General director Dr. ROBERTO MEIRELLES DE MIRANDE of the Brazil Agricultural Ministry, to Dr. SOSIGENES GOMES DA FONSECA the director of Institut de Pesquisa Agropecuâria do Nordoste together with his assistants who organised the study-tour in north east Brazil, helped prepare material and provided information as introduction to the various stations. I must also thank Mr. CLAUDIO CZAPSKI who accompanied me on the journey and acted as interpreter.

Because of these frequent declines in agriculture the economic development of the whole area is retarded. The Brazilian administration is particularly anxious to promote the development of the north-east.

Agricultural development is confronted with a double problem. The first question is: can modern arable-farming methods increase the average yield and prevent recurrent decline? The other question is: how and in what circumstances can crop production be increased by irrigation and by the introduction of different crop-plants? Large schemes have been proposed for irrigation and a few already completed. Although irrigation will lead to an important extension of the cultivated acreage and to an increase and diversification in crop production it will only provide a partial solution. In the following paper discussion will be of arable husbandry without irrigation, that is dry-farming. Irrigation will be considered in a later work.

In order to be able to asses which farming methods are appropriate and successful in different districts, it is necessary to know the present weather conditions. The Brazilian agricultural ministry and the Sudene[1] provided climatic means for various stations in the north-east, and in addition rainfall figures collected for other places by the hydrological division of Sudene. For the state Ceara information was available from 255 stations, for Rio Grande do Norte 78, Paraiba 76 and Pernambuco 82. Also the author was able to make a journey with an informed guide through the dry areas of Pernambuco, Ceara, Rio Grande do Norte and Paraiba. (Fig.1) The following statement is therefore in particular related to the dry zone of those four states.

Of the weather conditions temperature is astonishingly constant, high since it is a tropical region, with little variation in the whole year, although the range is a little greater at altitudes.

Of the 43 stations with data for the period 1931-60 the average annual temperature varied from 20.5°C at heights of 900 m to 28.0°C in favourable lowlands. The range between the warmest and the coldest months was in general 2°-4°. São Bento in Maranhâo state had however a range of only 0.9° (annual mean 26.4°), and Fortaleza only 1.2°C. The greatest range of 5.3°C was recorded at Monte Santo in Bahia state, at a height of 500 m, and with an annual mean of 23.4°C.

The absolute maxima are in general between 35° and 39° Celsius. Only in a few instances is 40° reached or exceeded.

The absolute minimum for most stations is considerably above 10° Celsius, only falling below that in the highlands. The absolute minimum at Ibipetuba on 27.7.51 was 6.1°, at Monte Santo 7.5° on 21.8.58, at Caetitá 7.6° on 29.7.37 and 25.7.49 and at Barra 8.8° on 27.7.51.

Ibipetuba had the greatest absolute range of the 43 stations: 34.2° Celsius. Quixeramobim had the lowest range of 17.5°.

The Sudene collected together in 1963 climatic data for north-east Brazil from 82 stations. The observations were not all for the same period but places were recorded for the period in which observations had been made.

In Olinda in the period 1911-42 the lowest absolute range was recorded. The absolute maximum was 33.9°, the absolute minimum 17.8°C, and the range 16.1°.

---

1 Sudene = Superintendência do desenvolvimento do Nordeste.

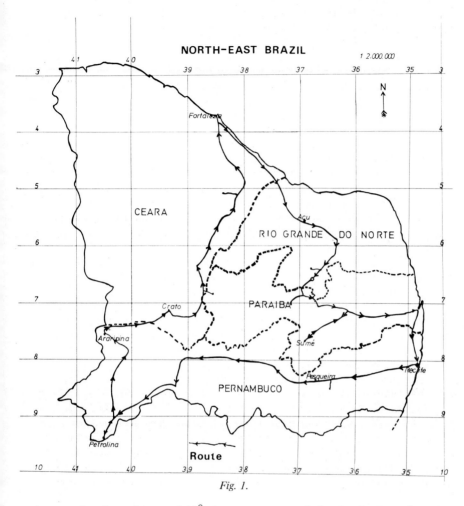

NORTH-EAST BRAZIL

1 2.000.000

CEARA

RIO GRANDE DO NORTE

PARAIBA

PERNAMBUCO

Route

Fig. 1.

An occasional maximum of 40° or over was recorded only six times from the 82 stations.

In terms of thermal conditions for the crop plants the temperatures are relatively stable the whole year and generally for warm-loving plants near optimum. The minimum temperatures for crop plants are approached only at altitude. It is possible however that temperatures in plants stands because of direct radiation, which can be very intense throughout north-west Brazil, may approach or even exceed the maximum for plants.

In terms of humidity and rainfall conditions to which the plants must adapt, the most important fact is the marked seasonality of north-east Brazil, a dry season and a rainy alternating. It is surprising however that neither of these seasons commence at the same time throughout the area, but are phased from north-west to east and then south. The hydrological division of Sudene has constructed a map showing the onset of the wet and dry seasons.

The rainy season begins in the south at Bahia in October and lasts to March.

103

Northwards, in a broad strip of Bahia, east Pernambuco, Piauii and Maranhao states, the rainy season extends from November to April. Still further north a narrow zone, not extending to the coast in the east, has rain from November to May. On the north coast the rains last generally from January to July, and in the north-east corner as far as Pernambuco from February to July. This applies to the coast only from Tauros to Joao Pessoa. From here to Salvador via the states of Alagoas and Sergipe the rainy season is delayed to March - August. In general the rainy season lasts six months. Only in Bahia state does the Sudene map show a small area with a rainy season November to January.

The phased beginning of the rainy season means that agricultural processes have to be related to its entry in order to make full use of the humid period. A staggered commencement of agricultural processes is possible because temperature conditions as described above in no way limit plant growth, and sufficient warmth is available the whole year to meet optimum conditions for plant growth. An exception may occur at altitudes with a rainfall from June to August when the temperatures are significantly lower. Similarly day-length varies little and does not induce any seasonal growth rhythm.

If only the average annual rainfall is considered then the values will be found astonishingly high for a dry area. The publication of the Sudene showed for 1963 that only three of 79 stations with rainfall data had an average annual fall of less than 500 mm. 28 had 500-1000 mm, 44 1000-2000, and four stations recorded over 2000. Similar circumstances are shown by the publication of the Agricultural Ministry for 1970. Of 43 stations only one, Remanso in Bahia state had just 500 mm. Fourteen had 600 to 1000 mm. 27 stations measured more than 1000, and three reached over 2000 mm.

In the mass of rainfall data assembled by the hydrological division of Sudene some 425 stations of relevance to our research were included. They did not all have figures for the same observation period, a number recording for the period 1911/13 - 1967, others from 1934/36 - 1967. Others have recorded for a shorter period but were included if the observation period was at least 20 years. A few stations had recordings for a long period, as for example Fortaleza 1849-1965, thus more than a hundred years. Quixeramobin began recording in 1896. At some stations measurement had suffered occasional breaks but though such breaks might affect special research projects but they are of little significance in this study.

The rank-size of the rainfall of the selected 425 recording stations is given in Table 2.

Only 27 stations have an average annual rainfall of less than 400 mm, and 2 under 300. These places are much subject to drought damage. A further 33 stations have rainfall of 400-500 mm. To a large extent these also belong to the areas subject to drought damage, altogether 60 stations, 14% of the total, whereas 71 stations, 17% of the total, have rainfalls of over 1000 mm.

In order to obtain a fuller picture of rainfall conditions in the whole area it was important to investigate the distribution of average annual rainfall amounts in the four states. Publications of the Sudene provided geographical co-ordinates for each station which were plotted on a map with isopleths at 201-300 mm, 301-500 mm 501-700 mm, 701-1000 mm, 1001-1600 mm and 1601-2200 mm (Fig.2).

The eastern coast and part of the western coast in the north have a high average rainfall of 1000 to 1600 mm. The very high rainfalls of 1600-2200 mm are present only in restricted areas. Where the northern coast bends westward the region with

Table 2. Average rainfall in mm.

	201-300	301-400	401-500	501-600	601-700	701-800	801-1000	1001-1400	over 1400
Ceara		1	7	19	42	56	44	38	9
Rio Grande do Norte		6	9	16	16	9	7	5	1
Paraiba	2	9	5	4	7	17	17	6	4
Pernambuco		9	12	18	10	4	8	3	5
total	2	25	33	57	75	86	76	52	19

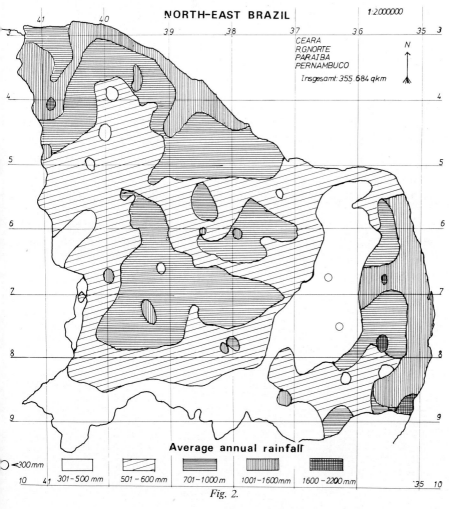

Fig. 2.

105

Table 3. The harvest in 1958 in comparison with the best harvest between 1960-70.

1958 harvest				best harvest 1960-70		
State	area	total t.	kg/ha	year	area	total t.	kg/ha
cotton							
Ceara	313,870	66,569	212	1961	500,077	208,795	418
R.G. d Norte	345,790	33,287	96	1960	388,086	119,793	309
Paraiba	331,960	73,720	222	1960	420,634	168,403	400
Pernambuco	343,946	65,517	191	1967	348,721	106,038	304
Total	1,335,572	239,093	178		1,657,518	603,029	364
beans (feijâo)							
Ceara	74,775	10,993	147	1967	348,760	208,288	598
R.G. d. Norte	51,113	16,910	301	1967	161,209	92,711	575
Paraiba	58,705	18,015	327	1967	183,226	110,926	605
Pernambuco	126,793	42,605	336	1967	260,391	155,660	598
Total	311,386	88,523	285		953,586	567,585	595
manioc							
Ceara	44,917	371,133	8,260	1967	84,993	1,368,799	16,105
R.G. d. Norte	25,558	124,417	4,868	1969	59,123	399,345	6,783
Paraiba	40,096	434,706	10,850	1967	64,532	695,474	10,777
Pernambuco	148,054	1,224,213	8,260	1970	140,597	1,644,323	11,660
Total	258,625	2,154,469	8,320		349,245	4,107,941	11,750
maize							
Ceara	78,460	15,667	198	1967	434,707	416,920	959
R.G. do Norte	50,441	14,406	286	1967	149,055	120,282	807
Paraiba	77,370	38,621	499	1967	251,981	230,666	915
Pernambuco	187,678	119,147	634	1967	325,609	307,611	945
Total	393,949	187,841	477		1,161,352	1,075,479	926

Table 3b. Planted area in 1958 as percentage of planted area in year of best harvest.

	Ceara	Rio Grande do Norte	Paraiba	Pernambuco	total
Cotton	62.6	88.8	78.8	98.6	80.6
Bean (Feijâo)	21.7	31.7	32.0	48.7	32.6
manioc	52.9	43.2	62.0	105.4	74.0
maize	18.0	33.8	30.7	57.7	33.9

501-700 mm reaches the sea. A little further south between longitudes $35°$ and $37°$W the driest region of all comes close to the coast and reaches as far at latitude $8°$. Here are situated both the stations with less than 300 mm. This area has a relatively small connection with the extensive dry-zone of south Pernambuco. The dry zone includes the large area with an average annual rainfall of 501-700 mm. Included within the zone are not only smaller areas of greater or lesser rainfall but in the centre between $37°$-$40°$W and between latitudes $6°$-$8°$ a large area with a rainfall of 701-1000 mm, forming the transition to the higher rainfall of the coast.

This cartographic presentation of annual rainfall shows clearly a few dry areas but provides no explanation for the fact that all of north-east Brazil is often subject to droughts that damage crop production.

To obtain further understanding it appeared necessary to compare rainfall distributions in one year with records of poor harvests. This was done for 1958.

The Bank of North-East Brazil (with its statistical handbook) made possible an evaluation of the reduced harvest of 1958, and further assisted by permitting use of unpublished data of plant production for the period 1960-1970.

In table 3 are the planted areas, the total production and the production per hectare for cotton, beans (feijão), maize and manioc in 1958 compared with the figures for the best harvest in the given state between 1960-1970.

In 1958 on average, for the four states considered, the yield of ginned cotton per unit area, of beans and of maize dropped to almost half that of a good harvest. Manioc was more succesful with well over 70%.

The poor harvest of 1958 is shown not only by the low yields per hectare but also by the reduced area sewn by the farmers owing to the failure of the rains. The area of cotton planted in 1958 was only 80.6% of that of a good year, manioc 74% beans 32.6% and maize 33.9% as an average for the four states.

As a consequence of the poor yield and the smaller planted area the total harvest of the states in 1958 was bad. Cotton was only 39.5% of a good harvest, beans 15.6%, manioc still 52.0% and maize 17.5%. It is to be noted that the figures are means, and districts are included that in 1958 had scarcely suffered aridity. In the true arid zone a total harvest-failure occurred. Where was the drought area and how extensive was it?

Table 4 shows the rainfall totals for 300 stations in Ceara, Rio Grande do

Table 4. No. of stations receiving the following rainfalls in 1958.

0-100	101-200	201-300	301-400	rain in mm 401-500	501-600	601-700	701-1000	over 1000
				Ceara				
14	34	46	28	18	9	2	3	1
				Rio Grande do Norte				
17	17	9	7	2	2	3	3	–
				Paraiba				
4	10	7	13	9	4	3	2	4
				Pernambuco				
4	3	14	11	5	4	3	5	7
				Total				
39	66	76	59	34	19	11	13	12

Norte, Paraiba and Pernambuco in 1958.

Of 329 places 39 or over 10% received less than 100 mm of rainfall. In total 181 stations recorded less than 300 mm, and only 36 more than 600 mm.

Table 5 shows how great was the rainfall deficiency in 1958. The percentage of stations in each rainfall class is shown for the average year and for 1958.

Table 5. Percentage of observation stations in different rainfall classes.
rain in mm

0-100	101-200	201-300	301-400	401-500	501-600	601-700	701-1000	over 1000
0	0		A) average annual rainfall					
0	0	0.5	5.9	7.8	13.4	17.6	38.1	16.7
			B) rainfall in 1958					
11.8	20.1	23.1	18.0	10.3	5.8	3.3	3.9	3.7

Whilst in the average year only 0.5% of the observation stations received less than 300 mm in 1958 55% were in this class. The percentage in higher rainfall classes is the reverse. In the average year 55% of stations had a rainfall over 700 mm, in 1958 only about 8%.

For agricultural considerations the table is not enough. It is important to know where the dry areas were situated and their extent. For this purpose the 1958 rainfall figures were plotted and then all the stations grouped according to rainfall classes 1-300 mm, 301-600 mm, 601-1000 mm, 1001-1500 mm, and over 1500 mm. In the first group with 1-300 mm rainfall not only was an outstandingly bad harvest recorded but also a considerable shrinkage of the planted area.

Fig.3 shows that the larger part of the whole district received less than 300 mm in 1958. In the south-west corner of Pernambuco near Petrolina one station recorded no rainfall for the whole year. In the quadrilateral bounded by longitudes 38 and 40, latitudes 6 and 8 most stations received in 1958 between 301-600 mm. This area generally has an average annual rainfall of 701-1000 mm (map 1). It is noteable that many places receiving higher than average annual rainfalls (1001-1600 mm) on map 1 appear on map 2 with reduced rainfall (601-1000 mm).

On the east coast a high rainfall was recorded in 1958, 1001-1500 and over 1500 mm. The north coast was almost completely in the dry zone with less than 300 mm. Only in the north-west between longitudes 38 and 40 where in normal years 1001-1600 mm is recorded, between 301-600 mm was received.

As a supplement to the rainfall data it was important to establish how many rainless and how many dry months (under 20 mm) were recorded at individual stations. In a dry period 20 mm of rainfall received in a month is of little significance for plant needs. Table 6 shows the number of stations that remained rainless for a definite period in 1958, or recorded months with rainfall of less than 20 mm.

Only one station of the whole 339 had no dry month. 6 stations had no month with more than 20 mm for the whole year. 58% of all observers reported 5 or more rainless months, 74% seven or more dry months.

The comparison of the rainfall totals for 1958 with those for the average year shows that most parts of north-east Brazil experience marked changes in rainfall totals from one year to the next. The irregularity in the rainfalls is probably the prime cause of uncertainty in the harvest in north-east Brazil, and must therefore

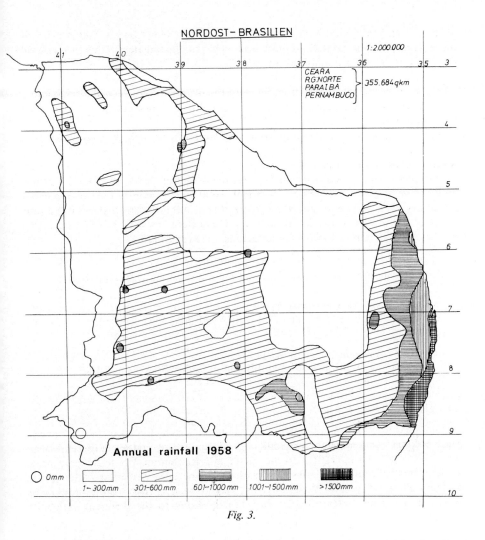

NORDOST – BRASILIEN

1:2.000.000

CEARA
R.G.NORTE
PARAIBA
PERNAMBUCO
355.684 qkm

Annual rainfall 1958

○ 0mm | 1–300mm | 301–600 mm | 601–1000 mm | 1001–1500mm | >1500mm

Fig. 3.

Table 6. 1958 number of stations remaining rainless for given months or receiving less than 20 mm in a given month.

month	0	1	2	3	4	5	6	7	8	9	10	11	12
number of stations rainless													
	21	17	17	31	56	45	53	45	23	21	7	2	1
number of stations with less than 20 mm													
	1	1	1	8	14	23	41	64	66	60	37	17	6

be quantified. As a measure of the range, after KÖPPEN, the difference of the maximum and minimum rainfall will be expressed as a percentage of the mean. KÖPPEN quoted for central Europe, for thirty year observation periods, 60% as normal. In Turkey in the dry zone of Inner Anatolia the rainfall totals show considerable fluctuation. At Urfa in south-east Anatolia the difference between the

109

maximum and minimum annual rainfall is 790.9 − 157.6 = 643.3, that is 140% of the average annual rainfall of 452 mm. This is the greatest value for rainfall fluctuations yet measured in Turkey.

In order to establish these values for north-east Brazil for each rainfall station with a long period of observation, likewise the difference between their maximum and minimum annual rainfalls must be expressed as a percentage of their mean annual rainfall.

For the minimum rainfalls the following were established: four stations had a completely rainless year, 73 had years with less than 100 mm, and 311 of 432 stations measured less than 300 mm.

The proportions were inverted for maximum rainfall years. Of the 432 stations only 49 or a good 10% had totals under 1000 mm. In contrast 97 places had a maximum over 2000 mm, and three more than 4000 mm. The highest in the four states as yet recorded was Sao Benedito in Ceara state at a height of 903 m, with a rainfall of 4583 mm in 1957. Further details are given in table 7.

Table 7. Number of places with maxima and minima rainfalls by class. Rainfall in mm.

A) maximum annual rainfall

501-601	601-800	801-1000	1001-1500	1501-2000	2001-3000	3001-4000	over 4000
1	14	34	162	124	76	18	3

B) minimum annual rainfall

0	1-100	101-200	201-300	301-400	401-500	501-700	701-1000	over 1000
4	73	128	106	50	32	25	13	1

The maximum and minimum rainfalls indicate to which class of rainfall range a place belongs. As a scale for variability the difference of rainfall in extreme years must be considered in relation to the annual average rainfall. Where the percentage value is small the rainfall shows little variation, where the percentage value is high the rainfall shows greater variation from year to year. Table 8 shows how many places in the four states of north-east Brazil belong to the different variability classes.

Table 8. Degree of variability in annual rainfalls: maximum − minimum expressed as a percentage of the mean.

80-100%	100-150%	150-200%	200-250%	250-350%	350-500%	over 500%
no of places 6	121	182	85	29	8	1

Only six places have a variability of 80-100%. The largest number, 182 stations, have a variability of 150-200%. Over 250% variability is shown by 39 stations. The highest variability was that of Bom Conselho in Pernambuco State. With a average annual rainfall of 480.3 mm, in 1944 2567.9 mm was recorded, in 1937 31.4 mm. This gives a difference of 2536.5 mm, 528% of the average annual rainfall.

110

It is further of importance for agriculture to know the spatial distribution of rainfall variability in order to know what variation may be expected in a given district (Fig.4).

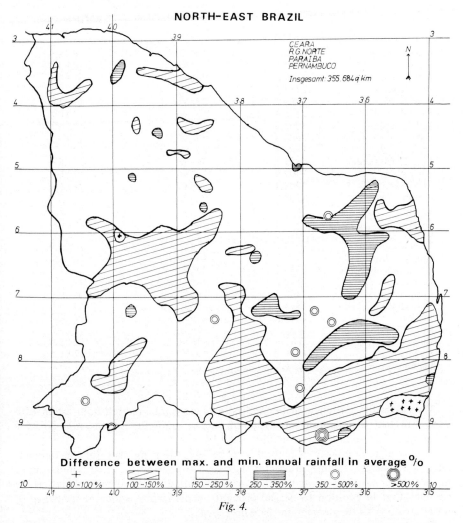

Fig. 4.

Fig.4 shows that in the four north-eastern states Ceara, Rio Grande do Norte, Paraiba and Pernambuco in the main the high variability of 150-250% applies. A few small districts in the whole area, situated in south-east Pernambuco between 35°-38° west, and 6°-7° north have a variability of 100-150%. The higher variabilities of 250-300% are locally distributed in the four states and in somewhat larger areas between 35°-36° west. The very high variability of 350-500% is limited to individual stations between 7°-9° S and 36-41° W.

As can be seen by comparison of maps 1 and 3, and as must be emphasised the high variability of annual rainfall is in no way limited to places with low rainfall, but is shown by stations with an average rainfall of over 1000 mm. (Table 9).

111

Table 9. High variability in association with high annual rainfalls

annual rainfall in mm.

place	average	maximum-minimum	difference as % of average
Cancuaretama	1320.7	2417.3	182.9%
Mamaguape	1534.7	2814.4	183.5%
Olho Dagua	1178.1	3689.8	313.0%
Alagoa Nova	1114.9	2747.1	246.2%
Bom Jardim	2159.1	3848.0	177.1%
Bieque	1175.4	1994.9	170.0%
Mata Grande	1033.3	1948.2	188.6%
Uniao dos Palmares	1004.9	1948.2	170.0%
Quebranculo	1801.9	3558.1	197.5%
Luiz do Quitunde	1581.5	3056.3	192.8%
Porto Calvo	1302.9	2218.5	170.0%
Luis Correia	1208.4	2941.3	243.5%
Piripiri	1702.3	3113.4	182.8%
União	1562.9	2961.9	189.5%
Acarau	1032.8	2085.3	202.0%
Granja	1018.6	1957.9	192.2%
Tucunduba	1013.7	2442.6	241.1%
Sao Benedito	1962.0	4092.1	208.5%
Guaraciaba do Norte	1248.8	2901.4	232.4%
Paracuru	1149.2	2672.0	232.2%
Fortaleza	1386.2	2632.5	189.9%
Maranguape	1344.8	2603.1	193.5%
Aquitaz	1312.8	2294.6	174.8%
Ar	1697.1	2965.4	174.6%
Guaiuba	1083.0	1986.6	183.5%
Mulungu	1131.4	1927.7	170.1%
Itapeim	1291.6	2248.5	173.9%

Thus even the station with the highest average annual rainfall Bom Jardin with 2159 mm had a maximum of 4458 mm and a minimum of 610 mm giving a difference of 3848 m and a variability of 177%.

In this high variability in rainfall is to be seen the main cause of the frequent failures of harvest, famine and water shortage suffered by the rural population of north-east Brazil. The relatively high average rainfall offers some explanation for the fact that in addition to typical drought phenomena considerable evidence of water erosion can be observed in the whole area.

The possibility of arable husbandry depends not on the isolated dry year or a particularly good year but much more on the question if a number of consecutive bad years occurs. In order to illustrate the conditions in north-east Brazil a graphical presentation of the annual rainfall at Paracuru, a coastal town north-west of Fortaleza, and at the research station Serido/Cruzeta in the dry area of Rio Grande do Norte is given (cf. figs 5 and 6).

The diagram for Paracura covers the period from 1912 to 1965, with the exception of the years 1934 and 1959 deficient in observations for a few months. The year 1964 is included even though observations were not made in July and September in the dry period. Since the rainfall for the remaining months was

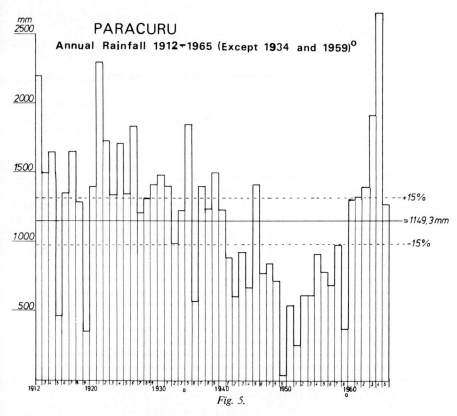

Fig. 5.

Research Station SERIDO
Annual rainfall 1930 – 1971

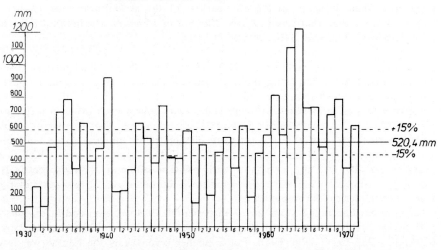

Fig. 6.

113

278.6 mm, the annual maximum, if rain did occur in July and September it would merely have increased the maximum.

As the fig. shows Paracuru between 1912 and 1932 had only the two years with very low rainfall. In the other years the totals exceeded the average of 1149.3. In order to emphasize marked deviations from the average the lines for + and −15% are added. Twice between 1912 and 1932 the annual rainfall exceeded 2000 mm. From 1941 to 1958 the annual total diminished considerably. Only once in this period was the mean clearly exceeded. In 1949 only 26.7 mm were received, the minimum annual total. In 1951 and in 1958 with totals of 242.3 mm and 361.8 mm there was still a considerable drought. After 1960 the rainfall totals increased sharply and, as mentioned, the maximum was achieved in 1964.

Fig.6 shows the yearly rainfall from 1930 to 1971 for the research station Serido. The average fall in this period was 520.4 mm. Most striking is the extremely dry period of three years, namely 1930-32 when the rainfall exceeded 200 mm only in 1931. A further period of aridity occurred in 1941 to 1943. In the remaining years the fluctuation in annual totals was relatively large. In 1940 more than 900 mm were received, and in 1963 and 1964 1000 mm exceeded. As in Serido the maximum occurred in 1964 with 1228.2 mm.

For arable farming it is of decisive importance if a grouping of dry years as at Serido 1930-1932, and 1941-1943 often occurs. It must also be noted that the occasional year with extremely high rainfall causes a disproportionate rise in the mean annual figure. At Serido the high rainfalls of 1963 and 1964 have the effect of lifting the mean from 487.4 to 520.4 mm. Further examples of the same effect can be quoted. Malhada Real in Pernambuco state received 1987.4 mm rainfall in 1960. If this year is ignored the annual mean sinks from 462.3 mm to 413.2 mm. Bom Conselho received 2567.4 mm in 1944. Without this maximum fall the mean would be 405. mm instead of 480.3 mm.

Important in relation to arable farming is heavy rain, as the result of which large quantities of water affect the soil in the short period of time. The publications of Sudene show seven places that have received more than 200 mm in 24 hours, and three more than 270 mm. In the publications of the agricultural ministry eight stations were named that received over 200 mm in 24 hours. The heaviest fall was at Maceio in Alagoas with 316.3 mm in 24 hours.

Generally throughout north-east Brazil heavy rain of high intensity can fall occasionally. 100 mm rain means 100 litre/m^2 or 1000 m^3/ha. Rain of over 100 mm in 24 hours can bring about considerable erosion even on gentle slopes, particularly if the soil is unprotected.

In order to ascertain how often drought years follow each other in different districts the stations with under 500 mm were examined. The number of years with less than 100 mm at a station were distinguished, similarly those between 101-300 mm, 801-1000 mm and over 1000 mm. The frequency with which dry years of less than 100 mm, and those with less than 300 mm, follow successively was also distinguished. The results are shown in table 10. A few particularly marked examples can be emphasised.

São Rafael had in 43 years an average rainfall of 451.1 mm and in the same period seven years with less than 100 mm a year, of which six occurred in two runs of three: 1952-1954, 1956-1958. For a further seven years the rainfall was less than 300 mm a year, and for two years it was more than 1000 mm. Cabaceiras is extremely dry. The average rainfall for a period of 54 years was 252.4 mm. Eleven

Table 10. Number of years with rainfall in mm class.

obs. years		1-100	*	101-300	*	801-1000	>1000	mean	variability
31	Hipolito	2		9	(4)	6	2	489.1	212.1%
33	Afonso Bezerra	1		7		2	2	471.1	232.4%
31	Queimadas	1		9	(4)	1	1	406.3	275.1%
29	Pixore de Baixo	3		9	(3)	3	1	423.2	375.0%
43	Sao Rafael	7	(3+3)	7		3	1	451.1	267.6%
32	Pedro Avelino	1		7	(4+3)	1	0	395.4	214.9%
50	Lajes	5		15	(5+4)	2	1	389.2	319.5%
31	Recanto	6	(5?)	6	(8?)	2	1	417.5	326.0%
31	Sao Tome	1		9	(3)	1	0	384.4	207.1%
51	Currais Novos	3		17	(6)	2	0	373.6	227.3%
40	Gargalheira	1		10	(3)	1	2	476.9	225.0%
53	Santan Cruz	0		9		3	2	489.3	300.0%
43	Serra Caida	0		9	(3)	1	2	498.6	341.3%
32	Equador	3		8	(6)	1	0	392.1	176.4%
31	Riacho Fundo	3		16	(4)	0	0	302.3	244.0%
51	Picui	1		26	(6+6)	1	0	313.0	248.4%
35	Barra de Sta Rosa	1		20	(7)	1	0	324.5	251.5%
27	Pedra Lavada	3		16	(12)	0	0	283.1	244.0%
33	Olivedo	0		8	(3)	1	1	465.5	260.0%
43	Desterro	6		18	(8)	1	0	321.8	288.8%
27	Salgadinho	1		7	(5)	0	0	417.8	153.4%
55	São João do Cariri	1		22	(4)	3	1	386.4	369.9%
51	Soledade	0		17	(7)	0	1	375.8	253.4%
32	Jofely	0		13		0	0	340.0	158.2%
44	Boa Vista	0		16	(6+5)	0	0	408.8	181.1%
54	Cabacieras	11	(6)	24	(15+7)	0	0	252.4	297.5%
30	São João do Tigre	0		12	(4)	1	2	413.6	269.1%
29	Sume	1		5		1	0	471.4	177.3%
35	Caraubas	0		15	(5)	1	1	359.6	264.1%
33	Bodocongo	7	(4)	6	(11)	1	0	333.8	293.0%
28	Arizona	0		11	(5)	0	0	383.6	144.5%
21	Pao Perro	2	(1=0)	10	(3)	0	0	344.8	217.0%
16	Santa Fe	3	(1=0)	3		1	0	348.1	276.5%
26	Malhada real	1		9		0	1	462.3	410.0%
53	Petrolina	0		18	(3)	1	1	396.2	216.4%
52	São Jose do Egito	5	(5)	15		2	2	427.1	242.1%
30	Betania	0		1		2	0	494.5	123.1%
27	Jeritaco	0		2		1	2	486.5	188.3%
27	Gravata	0		7		0	0	393.0	143.2%
54	Sta Maria de Boa Vista	1		15		4	2	459.2	219.2%
52	Cabrobo	0		12		3	1	469.8	223.5%
51	Floresta	0		14	(4)	4	0	456.1	173.5%
47	Belem São Francis	0		15		1	0	406.8	173.2%
31	Airi	0		5		1	0	464.0	159.8%
23	Sitio Novo	2		9		0	0	347.0	176.5%
25	Ico	0		7		0	0	401.1	119.0%
31	Moxoto	0		12		0	0	346.6	140.0%
27	Inaja	0		10	(3)	0	0	362.0	138.9%
25	Tara	2		7	(4)	0	0	392.7	176.5%
31	Petrolandia	0		10		1	0	407.3	162.5%
29	Bom Conselho	5	(4)	9	(7?)	3	1	480.3	527.5%
29	Delmiro Gouveia	0		6		2	1	497.4	163.1%
55	Piranhas	0		5		0	0	482.9	119.5%
34	Curatis	0		7		1	0	436.6	176.0%

115

Table 10 continued.

obs. years	1-100	*	101-300	*	801-1000	>1000	mean	variability
31 Jua	1		5		4	0	495.5	167.2 %
33 Santo Antonio	0		5		1	0	453.0	142.8 %
18 Ipuerinhas	0		9	(4)	0	0	316.4	148.3 %
33 Itagua	0		7		1	0	468.6	140.1 %
33 Poco de Pedra AC	0		6		2	0	488.2	139.6 %
33 Parambu	0		6		0	1	482.7	180.9 %

Of the 60 stations shown on the table, 35 have sequences of between 3 to 15 dry years

* = number of sequent years   Variability = maximum − minimum
rainfalls as % of mean

Table 11. Osmotic value at 20°C (W) for various relative humidities (h).

h%	W. Atm.	h%	W. Atm.	h%	W. Atm.
100	0	95	68.4	85	270
99.5	6.7	94	82.4	80	298
99	13.4	93	96.7	75	384
98.5	20.1	92	111	70	476
98	26.9	91	125	60	680
97.5	33.8	90	140	50	922
97	40.6	89	155	40	1190
96.5	47.5	88	171	30	1555
96	49.5	87	186	20	2055
95.5	61.4	86	201	10	2890

years had less than 100 mm and seven of these followed in sequence. 24 years had falls of 101-300 mm. One sequence of dry years lasted for fifteen years from 1945-1959. Since from 1959 to 1962 observations were not complete it is possible that the drought lasted longer. Another dry period, of seven years, lasted from 1917-1923.

The effects of dry periods are emphasized by low air-humidities. The lower the humidity the greater the energy with which the air absorbs moisture. The plants are subjected to rising transpiration. If the roots cannot take up moisture, wilting ensues and with longer periods of unfavourable conditions death of the plant.

Between the relative humidity and osmotic value there is a fixed relationship. WALTER gives the following table for this relationship:

In areas with marked rainy seasons the humidity is relatively high during this period, and falls considerably during the dry period. Terezina in Piaui state 5° 5' S, 42° 49' W, is an example and the average air humidity is shown against the average rainfall for the same month. (Table 12).

The table is completed by including a place with a relatively higher and steady air humidity, Sao Francesco do Conde in Bahia state 12° 37' S, 38° 40' W, and a place with a very low air humidity, Remanso in Bahia state 9° 41' S, 42° 4' W.

In Terezina the relative humidity in March, the peak of the rainy season, was 85.2%. It fell in September, the middle of the dry season, to 55.2%, a fall of 30%.

Table 12. Average monthly rainfall and mean air humidity.

Month	I	II	III	IV	V	VI	VII	VIII	IX	X	XI	XII

A) Terezina Piaui, 5°5' S, 42°49' W, height 49 m, rainfall in year 1291 mm

	I	II	III	IV	V	VI	VII	VIII	IX	X	XI	XII
mm:	174.7	235.5	310.8	253.6	90.7	12.9	7.8	6.2	9.5	27.7	60.2	105.4
%	76.8	82.7	85.2	82.5	81.2	73.6	65.2	57.2	55.2	56.9	61.4	68.4

B) Sâo Francisco do Conde 12°37' S, 38°40' W, height 80 m, rainfall in year 1713.9 mm

	I	II	III	IV	V	VI	VII	VIII	IX	X	XI	XII
mm:	76.6	78.0	163.2	289.5	298.2	193.4	206.0	110.7	85.2	94.6	140.9	98.5
% :	82.1	82.1	84.2	86.7	88.6	88.5	89.1	87.4	86.0	84.1	83.3	82.5

C) Remanso Bahia 9°41' S, 42°4' W, height 411 m, rainfall in year 496.7 mm.

	I	II	III	IV	V	VI	VII	VIII	IX	X	XI	XII
mm:	91.8	64.7	107.9	34.7	10.4	1.2	1.4	0.0	3.7	10.1	77.7	93.2
% :	50.5	51.5	51.4	51.0	49.8	49.9	49.9	48.4	46.3	45.0	46.7	48.4

A similar variation in air humidity must be seen as characteristic for all areas of north-east Brazil having seasonal rainy and dry periods.

Conditions are quite different at Sâo Francisco do Conde in Bahia. All months have recorded rain over 70 mm for the 30 year period of observation. The humidity in July averaged 89.1%, and sinks in January and February only to 82.1%. The difference between maximum and minimum relative humidity is only 7%.

An extremely low air humidity is shown by Remanso, Bahia, in all months. As can be seen from the table in the averages for 30 years the highest average humidity for February was 51.5%. The figure for September was 45.0%. The annual mean was 49.1, 6% beneath the minimum value for September for Terezina. The relative humidity at Remanso is as low for the whole year as it is in central Anatolia in summer (Table 13). Remanso, in view of the high temperature of 37.1° at the middle of the year and the extraordinarily low humidity, and despite its average rainfall of almost 500 mm, belongs to a region of extreme aridity.

Table 13. Average relative humidity in summer in central Anatolia.

region	May	June	July	August	September	October
Ankara	57%	51%	43%	41%	46%	56%
konya	56%	50%	41%	40%	46%	48%

The effects in a dry year of increasing drought and decreasing air humidity can be dangerous. Unfortunately data for the changes in humidity for north-east Brazil during the extremely dry period of 1958 are not available. I am grateful that the meteorologist of Ipeane supplied figures for various years between 1933 and 1967 for Cruzeta. 1933 followed three markedly dry years with rainfalls of 373.5 mm, 253.6 mm, and 100.0 mm. 1933 itself received 644.9 mm, but 453 mm of this total fell in February, March and April. The remaining months were markedly dry. The relative humidity rose only in April.

Heavy rain fell from December 1963 to May 1964. In June and July the rains continued. Even August and September received some rain. Only October to December were dry.

Table 14. Cruzeta R.G. Norte: average monthly air humidity 1933 & 1964.

months	I	II	III	IV	V	VI	VII	VIII	IX	X	XI	XII
1933:	60%	64%	63%	73%	60%	58%	59%	53%	53%	53%	50%	53%
1964:	73%	74%	77%	78%	79%	72%	70%	63%	62%	55%	55%	53%
difference	13%	10%	14%	5%	19%	14%	11%	10%	9%	2%	5%	0%

If the above table is compared with table 11, the osmotic values at various relative humidities, then it becomes obvious how high the water loss of plants was in 1933 compared with 1964.

Unfortunately the average value of relative humidity does not give a full picture of the conditions affecting transpiration.

During the daylight, particularly when the sun is shining, relative humidity falls. This is shown by table 15, with relative humidities at three times of day, based on thirty years observations. Also shown are the absolute minima recorded during the same period of observation.

Table 15. Ankara. Average relative humidity 30 year means.

month	7h	14h	21h	Absolute minimum
March	80%	50%	67%	5%
April	73%	41%	59%	6%
May	71%	40%	61%	12%
June	64%	34%	53%	5%
July	57%	28%	43%	3%
August	58%	26%	40%	7%
September	63%	30%	46%	7%

The average relative humidity in Ankara falls in the summer months from seven in the morning to midday and to 14.000 hours by a good 30% and then in the evening to 21.00 hours rises again on average by 17%. This drying-out of the plants and soil due to low humidities would be still further advanced by the occasional very low values. In many parts of north-east Brazil during the dry season similar conditions prevail.

The importance of relative humidity for the water budget of plants stems from the fact that air becoming drier increases its evaporative effect and leads to increased water loss from the plant. If the roots are unable to provide the water from the dried-out soil then the plant dries out.

All the factors influencing the water budget of plants are subject to great variation in north-east Brazil. For agriculture a full picture of the average climate and the extreme weather conditions is wanted. The data should be presented in simple form, or in an easily comprehensible outline as made possible by temperature and rainfall measurements carried out even by small weather stations.

Simple forms of presentation are Lang's rain factor, de Martonne's aridity index and Köppen's climatic formula. The last has the advantage that is can be extended and indicate unusual weather conditions associated with some climates.

Flexible and giving an overall view is the climatic diagram developed by WALTER. The monthly precipitation totals and the average monthly temperatures are shown. The scale of the diagram is so chosen that 10° Celsius corresponds with 20 mm of rain. The individual monthly values are connected by a curve. If the precipitation curve oversteps the temperature curve humid weather is indicated. If on the other hand the temperature curve oversteps the precipitation curve a dry period is indicated. For clarity the moist periods are marked with vertical shading, the dry are dotted. Above the diagram the name and altitude of the site is given. On the left-hand column are the mean annual precipitation amounts, on the right-hand column the annual average temperatures. In the same way the weather of an individual year can be shown. For the presentation of such diagrams only a knowledge of precipitation and temperature conditions is required, and this is available for many stations. The basis of the climatic diagram is the relationship of temperature and evaporation. The temperature and the evaporation curves are parallel in many dry areas.

In Fig.7 climatic and weather diagrams are given for 1930, 1931, 1932 and 1964 for the research station Serido. Since the temperature data of the station were not available those for Iguatu were selected, situated in the same latitude and at the same altitude. A certain departure of the temperature from average values, above for the dry years and below for the wet years like 1964, is accepted. The characteristic statement of the diagrams is however not essentially influenced by these inexactitudes.

The climatic diagram including the period 1930-71 shows a well marked moist period beginning in February and ending in May. The dry period extending to January is very clearly marked.

The four weather diagrams show scarcely any relationship with the climatic diagrams of the same stations. 1930 and 1931 had only a very meagre moist period in March-April. All the remaining months were very dry. In 1932 the drought lasted the whole year. In 1930 and 1932 scarcely a quarter of the average rainfall was received, and in 1931 only a half.

A completely different picture is shown by the weather diagram for 1964. The moist season begins in January and brought well-above average rains, in February, March and April above 200 mm. The dry period did not begin until August, but only in the two last months were rains negligible.

A consideration of the climatic and weather diagrams reveals even more clearly than fig. 6, the total annual rainfall 1930-1971, that the variations in rainfall in individual years are sufficiently large and frequent to affect the mean and the usefulness of the climatic diagram.

From the view point of arable husbandry the weather and climatic conditions in north-east Brazil are favourable in terms of temperatures throughout. Thus the occupier of a facenda five kilometers from Maria de Boa Vista in east Pernambuco can with care and by means of irrigation provide fresh grapes each week of the year. A trial plot of lucerne with irrigation and under these favourable temperature conditions gave eleven cuts, whilst in southern Europe only about five and in central Europe 3-4 are taken.

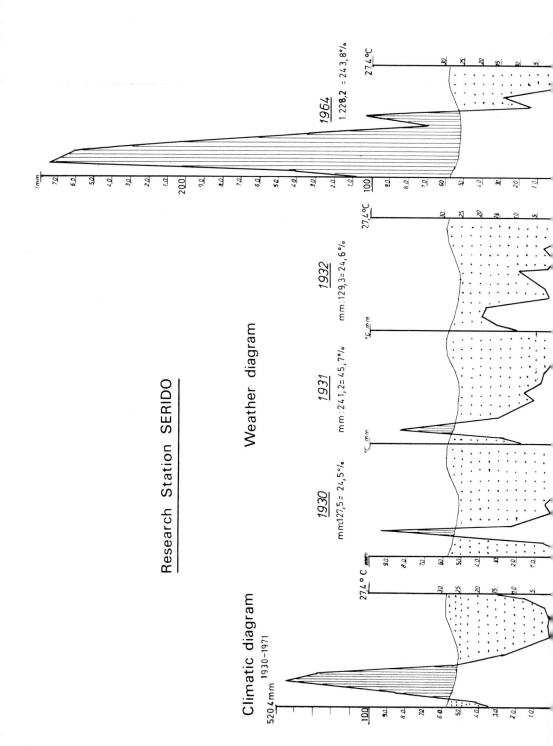

Research Station SERIDO

Weather diagram

Climatic diagram
1930-1971

Made difficult and from place to place damaged, arable farming in north-east Brazil suffers from year to year over practically the whole area great variations in rainfall. Often a moist year is followed by an extremely dry year. In regions with low average rainfalls several dry years can follow in sequence. As a result there is not only a series of bad harvests but also a lack of fodder and difficulties with water supply for man and beast. Again and again a dry year occurs, as in 1958, that affects the larger part of north-east Brazil.

The particular weather conditions of north-east Brazil present agriculture with a question, to what extent and where is it possible, in the light of the most recent knowledge of dry farming, to achieve an average yield sufficient to guarantee farmers a viable economy?

The productive plant growth is determined by the precipitation of the rainy season. A pre-requisite for success in cultivation is therefore a well-structured soil, with high water capacity and a good depth able to absorb moisture and store it for plants at a reachable depth.

A particular difficulty for the whole region is the marked affect on soil capability due to the weather. This applies in both the wet and the dry season. In the wet season heavy rain by breaking the soil granules and washing away fine particles leads to soil erosion. In the dry season wind erosion occurs and fine particles are blown away. Before the farming methods are described attention must be paid to the various forms of soil erosion and the possibility of combatting it.

Heavy rain is in no way limited to areas with a high annual rainfall. It can occur also in districts with low rainfall totals. Of the sixty stations in four states with an average annual rainfall of 250-500 mm the following monthly totals were observed.
Number of stations with monthly maximum rainfalls:

of	200-300	300-400	400-500	500-600	600-700	700-1000	>1000 mm
	6	28	18	3	3	1	1

It is completely possible that, with a monthly fall of over 200 mm, more or less heavy rain can occur. In the climatic data assembled by the agricultural ministry and Sudene figures are available of maximum rainfall in 24 hours for 123 stations. As shown by the following table only three stations had less than 85 mm as their maximum for 24 hours. In no less then 44 stations the maximum was over 150 mm.
Number of stations with observed heavy rain in 24 hours:

of 64-84	85-100	100-150	150-200	200-250	250-300	>300 mm
3	22	54	28	9	6	1

As an example of the occasional particularly-heavy rainfall, in an area which is part of the rain-deficient north-east Brazil, Malhada Real can be mentioned. It is situated in Permambuco state at 9°;' S and 40°31' W, at an altitude of 345 m. The average annual rainfall in 34 years was 462.3 mm. In March 1964 in one month 1377.2 mm were measured.

A similar deluge must have been suffered by Olho Dagua in Paraiba state in

121

1964 and 1967. The place is at an altitude of 275 m., and situated 7°13'S, 37°46'W. In 1964 in March alone 1700 mm were received, and in April an additional 958.2 mm. Since 398.9 mm had fallen in January and 397.7 mm in February, in only four months 3454.8 mm fell. In part this must have been made up of extraordinary heavy rain. To visualise the quantity of water involved it is necessary to realise that in the period each square metre of the district received 3455 litres, or each hectare 34,548 m³. In March 1967 Olho Dagua received 1197.6 mm, and in the four months February to May 2805.6 mm. In contrast in 1936 the rainy season gave only 72.5 mm.

It must be noted that details of the rainfall amounts in 24 hours do not give a clear picture of rainfall intensity, the rainfall received in a given time interval, that finally decides the erosive power of a rain storm. This can be illustrated by considering two heavy rainfalls at Cölasan, that produced similar quantities in 24 hours. The rain commenced on the Black Sea coast. That at Samsun measured 6.6 mm in the first minute, on average for the first five minutes 1.5 mm, in the first ten on average 1.38 mm, and averaged for the first half hour 0.61 mm per minute. At Kastamonu 5 mm fell in the first minute, on average 2 mm for the first five minutes, 1.6 mm for the first ten, and for the whole hour on average 1.14 mm per minute. In Kastamonu 65.3% of the total heavy rainfall occured in the first hour whereas Samsum received only 19.4%. Fig.8 shows clearly the comparison of differing intensities of heavy rain.

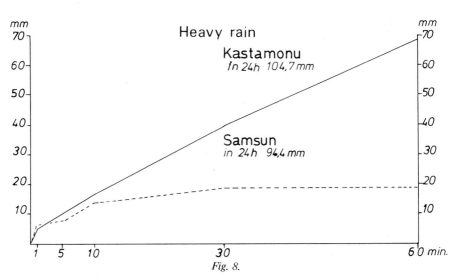

Fig. 8.

In order to produce erosion the size of rain drop is important. The larger and heavier the rain drop the greater the power to break the soil crumb and to remove the fines. With a sloping soil-surface, precipitation produces sheet erosion, the upper surface is washed away. If there are depressions in the direction of slope or plough furrows the water flows into them, deepening them to rills. If the heavy rain lasts or reoccurs the rills are deepened to gullies.

On steep slopes where the protective plant cover has been removed by felling or by overgrazing persistent rain can lead to a movement of the whole soil, to slipping (Fig. 9).

Fig. 9. Slips of a forest soil on medium slopes as a result of forest removal and soil saturation.

If the sloping land is not protected against water erosion the soil will be robbed of its fine particles and of most nutrients. It becomes infertile and at the same time loses its water retentive capabilities, leading therefore to aridity. In hilly land the soil removed by erosion will be deposited at the foot of the slope. On steeper slopes the fines moves to the streams and rivers.

If the arable land on slopes is not to lose its fertility, or be completely destroyed, the farmer must combat erosion.

The second danger to the fertility of the soils of north-east Brazil is wind erosion, the blowing away of soil fines from the surface in storms during the dry season.

As shown above the dry season in almost all parts of inner north-east Brazil can occasionally be extended beyond the average. On such occasions the soil dries out to great depths. If not protected by vegetation or by mulch the soil can then be blown away in storms, the wind becoming laden with fine particles and acting like a sand blast. Any surviving plants are torn to pieces and further soil particles broken up.

In contrast to water erosion that of the wind is more difficult to measure. In addition to the destruction of any vegetation present, wind erosion lessens the fertility and the water retentive capabilities of the soil. The valuable soil fragments are removed, and water capacity reduced, endangering subsequent cultivation with an increased drought risk. At the same time the soil loses its ability to absorb nutrients, and thus becomes infertile.

Wind measurements that would allow some evaluation of the wind erosion for individual districts of north-east Brazil are unfortunately isolated. The Sudene in

123

Fig. 10. Cotton on a steep field. On the steep slope the soil is eroded and the cotton dried out.

Fig. 11. The same field as in Fig. 10. At the foot of the slope on level ground the eroded soil is deposited and the cotton does well.

Fig. 12. Wheat sewing destroyed in a dust storm; Inner Anatolia 1962.

Fig. 13. A deep road ditch filled with fine soil; Inner Anatolia dust storm (6.4.1956).

its collection of climatic data in 1963 did list winds above force 8, thus having a storm character.

In Pesqeira in Pernambuco state on average at least one storm a month was observed. In the dry season the number rose to three or four. In umbuzeiro in Paraiba state on average for the years 1929-42 four storms were recorded in September and in October, three in July and in each of the months from October to February, and in May two. The largest number, 55 storms in a year, was recorded at Lençoes in Bahia state 12°55'S, 41°32'W, and at an altitude of 560 m, as an average 1932-42. The minimum per month was three, recorded in June and December, four in March, in April and in July and in August and in November, six in May and no less than seven in January.

It is unquestionable that wind erosion, the blowing away of the soil, in normal dry years but particularly in drought years, is a serious hazard to soil fertility and water retention over a wide area of north-east Brazil.

In an area in which the soil capability is at risk from wind and water erosion, the aim of the farmer must be to counter any degeneration in the soil, to maintain fertility and water retention, the decisive factor in the dry period for the success of the harvest.

To combat water erosion on slight slopes the soil should be prepared at a slight angle following the contour. Where slopes are greater small embankments should be built along the slope not to hinder the preparation of the land but to hinder the movement of rain water and therefore of soil movement. On still-steeper slopes the embankment should be accompanied by a furrow leading to a ditch. With heavy rain the embankment stems the flow of water which is led by the furrow into the

126

ditch, there to soak away. In order that it should not be washed out the slope of the furrow should not be more than 0.5%.

If the slope is still steeper arable farming must remain below. The land, if not forested, must remain under a protective plant cover. It is possible where it appears desirable to clear the natural flora in strips and to sow drought-resistant plants. This should only be done in small strips along the contour to avoid soil-loss in heavy rain. The newly sown strips should alternate with strips on which the original flora remains or which were previously sown.

To protect the soil from wind-blow, strip cultivation is effective. Perennial cotton can be alternated with legumes, maize or manioc in strips at right angles to wind direction.

If the danger of wind erosion is great, wind breaks must be planted at right angles to the main wind direction. Well suited for this purpose in north-east Brazil is *Euphorbia tirucalli*. This was often seen in well developed hedges, as for example in the cotton nursery at the research station Serra Talhada (Fig.14). Low bushes must be planted on both sides of the hedge to break the wind at ground level when the euphorb has grown tall. A well-sited wind break not only protects against wind erosion but also creates between the hedges, if they are not too far apart, a favourable microclimate for plants.

The farmer must also ensure that when slow growing crops are planted the soil should not remain long without a vegetation cover. It is to be approved when between newly sown cotton for the first year gourds, beans or other legumes and maize are planted. This gives protection against both wind and water erosion.

Fig. 14. Wind-break of Euphorbia tirucalli at research station Serva Talhada.

An important means of soil protection is mulching, that is covering the ground with plant remains. This can be chopped weeds or stubble, or other chopped plant

127

debris. Mulching has a favourable effect on the soil structure, assisting the penetration of water, and at the same time inhibiting evaporation. Also a soil covered with a mulch is less damaged by heavy rain. The impact of the rain drops is broken and splash erosion prevented. The flow of water is slowed, and also reduced by loss into the soil. Finally a mulch also protects against wind-blow. Mulching is therefore of particular importance in the protection of soil.

Investigations by CHEPIL (1958) showed how wind erosion is affected by different methods of handling the soil surface. He provided figures for the amount of fine earth blown from the soil in a given period. The storm CHEPIL observed had a speed of 10 m per second at 30 cm above the surface. Immediately at the surface, with a level prepared surface, the speed was still something above 7 m per second. If the level surface was mulched with wheat straw the speed at the surface dropped to 5 m per second. If the soil was laid in furrows at right angles to wind direction the speed sank to 3 m per second, and if the furrows were mulched the speed was lessened by a further metre per second. CHEPIL also reported that the storms removed from a level surface over 2000 kg. of fine earth per hectare in five minutes. With a mulch on a level surface the loss was 334 kg per hectare. On the furrowed parcel the loss was higher, 554 kg per hectare, but when mulched reduced to 125 kg per hectare. The effect of wind erosion was thus reduced, compared with the smooth surface to 6.25%.

The selection of a rotation must also bear in mind the problem of erosion. Stripplanting has an increased significance. This applies to both wind and water erosion. Well suited in the rotation are such fodder crops as lucerne, and esparsette. In Turkey there are forms of lucerne and esparselte with a higher resistance to drought. By careful husbandry a good ground cover can be maintained the whole year, protecting the soil against wind blow and against heavy rain. With strip cultivation the plants break the strength of the run-off and lessen the possibility of soil wash, and of wind blow.

How different plant stands, rotations and systems of cultivation all influence water erosion was shown by Musgrave in the U.S.A. and by Dr. MOACIP DE FREITAS in Pesqueira.

MUSGRAVE tested water erosion with various plant stands on the north Pacific coast of the U.S.A. The highest soil loss of 76 t. per hectare he valued as 100. For various plant stands he established the following relative values.

meadow land under humid climatic conditions or irrigated and forest	0.001
lucerne	5
fruit and vines with intercropping	10
fruit and vines with bare soil	90
fallow with wheat stubble	60
row cultivation (maize, cotton, tobacco, potatoes)	100
open fallow	100

The highest removal of 76 t per hectare corresponds with an annual loss of 3 mm.

Dr. MOACIP DE FREITAS measured water run-off and soil loss on arable land with a 12% slope at Pesqueira. The results established by him are shown in a table below. They show that even on relatively steep slopes water erosion can be inhibited.

Attempt by Dr. MOACIP DE FREITAS to measure soil erosion 1951-58.

rotation	preparation	water m³	run-off %	soil kg	loss %
tomato – legume – tomato	plough	9.491	100.0	413.9	100.0
tomato – tomato – tomato	mulch	7.516	79.2	191.5	46.2
tomato – legume – tomato	mulch	2.743	28.9	55.6	13.4
tomato – steppe – tomato	mulch	927	9.8	10.2	2.5
tomato with compost – steppe tomato with compost	mulch	681	7.7	14.1	3.4
tomato – steppe – steppe – tomato	mulch	511	5.3	5.3	1.3

The results show that mulching reduced the water run-off by almost 30% and the soil loss by 13.4% with the same rotation: tomato, legume, tomato.

The smallest loss of soil and the smallest run-off was achieved by Dr. FREITAS when he followed mulched tomatoes with two years of steppe, the native flora, and then a further year with mulched tomatoes. The water loss was then a little over 5% and the soil loss a little over 1%. With composted tomatoes, steppe, composted tomatoes the run-off, even when mulched, was almost 8%, a little more, and the soil loss was over 3% but still sustainable. The average yield per areal unit is however more considerable.

The research established that a slope of 12% when ploughed and cultivated normally is subject to too great a loss. It also demonstrated that with a good soil cover of plants, and by mulching, the water run-off and soil loss can be so reduced that the soil capabilities are in now way impared.

It was necessary to make clear the danger to arable farming under the special weather conditions of north-east Brazil, and the means of maintaining fertility, the difficult task upon which successful arable farming in north-east Brazil is dependent.

If the farmer of north-east Brazil wishes to achieve a useful harvest in years with poor rainfall he must ensure that his soil takes up as much rain as possible and stores it so that it may be reached by the roots of cultivated plants. At the same time he must protect his water reserves from loss.

Conditions for success are good deep-soiled arable, purposeful and timely cultivation, a careful selection of the crops plants and of the rotation, application of fertilisers at the right season, careful protection of the soil and regular weeding since the wild plants are, for water and nutrients, the successful concurrents of the crop plants.

The essential conditions for successful cultivation in dry areas is a correct choice of a suitable soil. It must be deep and of good structure so that it can absorb nutrients, and have a good water capacity. The following example shows that on such soils, even in arid areas, successful cultivation is possible.

In the Negev desert in the south of Israel EVENARI has shown and practically demonstrated that successful arable husbandry was carried on before our era, with an average rainfall of just 100 mm. The necessary condition was the presence of loess on the slopes and in the valleys. The depth of loess in the valley fields had to be 1.5-3.0 meters. With the onset of heavy rain, which is rare but nevertheless occurs most years, the loess slopes were covered with an impermeable layer and the wave of rain water led into the valley on to the arable. The soil was saturated

129

to depths. With this reserve of water it was possible to produce a crop. EVENARI produced by such means on a farm near Avdat in 1964 1800-2400 kg of barley per hectare and 2500-3000 kg of wheat per hectare.

In Egypt the single saturation of the arable fields by the Nile high water was for thousands of years the basis of agriculture. Possibly the same system was employed, with a dam, in the kingdom of Saba near Marib. It is also employed in the Yemen on the coast with streams coming from the interior. In these methods the soil acts as a reservoir, filled once for a crop.

Dry farming also depends upon the possibility of storing water in the soil for a long period. By having a year of bare fallow, the soil without plant growth, an attempt to conserve the larger part of the rainfall to be available to the plants in the planting year, together with any further rain. The degree of success that can be reached is shown in Fig.15. The seed planted after the fallow year found enough moisture in the soil to germinate and grow. The well developed seed has not yet received rain.

Fig. 15. East Anatolia. Wheat planted by seed-plough growing well after the fallow year although no rain had been received.

The first problem that arable farming in north-east Brazil has to solve, in relation to its dry season and possible drought years, is when and how to prepare the land to give a good seed bed but also to take up rainfall quickly and to protect from evaporation.

As shown above, in a normal year in north-east Brazil rainy and dry seasons alternate. The annual beginning of the rainy season, as the above described researches of Sudene show, different in different districts. Since the thermal conditions vary little the whole year, it is the onset of the rains or the beginning of the dry season which are the criteria for the farming year and not a definite calendar month.

130

Fig. 16. 'Graham - plough' = a goose-foot cultivator with harrow breaking up stubble.

Fig. 17. Graham plough.

The question of the method of soil working and seed bed preparation may be discussed for dry areas with 600 mm rain. Heavy rain occasionally occurs even in these regions but there is scarcely any eluvation of fines or any accumulation of fines and nutrients at depth. The turn-plough is therefore not required to bring the illuviated material to the surface, or to break up a compacted zone. It is sufficient to open up the soil to the required depth in order to increase the intake of water, and to ease penetration by roots. At the same time weeds must be destroyed.

In the U.S.A. a grubber with an attached spade harrow has been developed. This apparatus, the grubber having strong goose-feet, can work the soil well at depth, a necessary preparation in dry areas (Fig.16 and 17).

On sloping sites the plough must be employed along the contours to throw up low embankments as a protection against soil erosion. The grubber can be employed between embankments but it too must follow the contours.

The second question is the period of soil preparation giving the best results in water storage and in the conservation of fertility. With early sewings of annuals such as millet, maize, beans etc., according to the duration and amount of the average rains, some moisture can be left in the ground after harvest. Immediately after harvest the soil must be worked to prevent loss by evaporation and by weeds. Some moisture is then preserved for the next crop. The disadvantage is that with moisture and warmth the soil fauna is active and humus is decomposed. In a dry soil this activity is less.

FINCK (1963) suggested the following points in relation to the bearing capability of tropical soils and humus loss.

A. Positive effects

1. release of nutrients from the humus during mineralisation
2. Increased mobilisation of nutrients from the mineral reserve due to increased production of acids and chelates.
3. Intensification of soil-life during break-down.

B. Negative effects

1. lessening of sorption capacity (ability to retain nutrients)
2. lessening of water capacity (ability to retain moisture)
3. lessening of structural stability
4. lessening of resistance to erosion
5. reduction of soil life after break-down

For north-east Brazil the reduction of water capacity and of the resistance to erosion is particularly unfavourable.

A further disadvantage of the early preparation of the soil is the proneness of the uncovered soil to wind erosion during the dry season. These conditions are quite different when the soil during preparation is covered with a mulch. For this a grubber with a trailer draw-harrow is well designed. The grubber pulls out the stubble and weeds which are sliced by the harrow and strewn as mulch. The land is to a considerable degree protected against wind blowing, and moisture is preserved. A further advantage is that water from any isolated showers is rapidly absorbed.

Later the mulch is worked into the soil and adds to the humus. How far it is possible in north-east Brazil to maintain soil structure, fertility and water capacity by these means for individual soil types will only be ascertained by long field trials.

Another possible time for the preparation of the land for seed is the end of the dry period, beginning of the rainy season. In this case also the stubble and weeds are uprooted and cut for a mulch by a grubber. With the beginning of the dry period the ground will quickly dry out since the wild plants will rapidly use up any remaining moisture. In the dry soil the activity of the soil fauna will be inhibited and humus break-down slowed. The disadvantage is that during early growth the crop is dependent upon the rains.

If soil preparation at the end of the dry period or the beginning of the wet period is accepted the seed should be sown immediately. Recently a shredder with roller and drill has been developed. The shredder lossens the soil and breaks up the stubble and weeds. The light roller presses the loose soil a little, and at the same time the released seed. Preparation and sowing take place together.

WALTER C. DE SOUZA, agricultural adviser in Paraiba state reported to us in Santa Lucia that upon his initiative cotton fields were worked in the following way. In the hitherto unused fields furrows were drawn at the required distance apart for perennial cotton and the seed laid directly in the furrows. After the first rain the cotton grew quickly. The weeds between the rows can be removed with a harrow as soon as the cotton is established. Although cotton fields were a little further from Santa Lucia and in contrast to those nearer the town, had not received in July 130 mm rain, they were better developed.

This method of sowing cotton seems worthy of further testing. A complete preparation of the field is not required for the furrows are placed the correct distance apart, and the seed can be sown promptly. The furrows take up any water flow, favouring the germination, rooting and first development of the young plants. The remainder of the field, until the hoe is used, is protected by wild plants.

In most cultivation the preparation of the soil and seed bed is separated from the sewing, and in these cases the question of the purposeful completion of the work must be answered. First it must be emphasized that on gentle slopes to hinder water erosion, the seed rows should run along the contours. This is particularly important when perennial crops are involved. Where on level land, wind erosion is feared, the plant rows should be at right angles to wind direction.

Also of particular importance is the time of sowings. The aim in all dry areas should be that the crop to be planted uses to the full rainy season. Late sewing or planting, long after the beginning of the rains, is always associated with a great risk of harvest failure.

If some soil moisture is preserved by working the soil by mulching this can best be used for seed that for germination requires comparatively little moisture. The seeds to which this applies are shown by the work of KARL SILBERSCHMIDT and his associates in the biological institute of São Paulo. They investigated moisture required to germinate seeds of crop plants. In the discussion of the results it is stated 'If we select a few examples of seeds with low, medium and high substrate moisture requirements we could include in the first group broccoli, eggplant and corn, in the second bean, pumkin, water melon, rice and sunflower and in the last catjang, lab-lab and peanut.

In a light moist soil the seed is best placed on the floor of the furrow in order to come in contact with the moist earth, making germination possible. Some drier soil should be laid on the seed in order to hinder evaporation. The seed can then germinate without rain and develop. The first rain stimulates rapid growth of the plant and even in a dry year a worthwhile yield can be produced.

The rotation should also play a part in combatting aridity. By the rotation the water needs of the sequent crops can be favourably influenced. Thus many legumes are deep rooting on good soils, for example lucerne, esparsette, lab-lab and others. The following crop can use the root tubes of the legume, and by such means drive their own roots to greater depths. The water provision is improved in that a greater depth of soil is established. The benefit of the leguminous first crop is increased by the soil enrichment due to the nitrogen fixation of the symbiotic bacteria in the root nodules.

Reference may be made here to the success in the dry zone of Central Anatolia due to the introduction by ÖMER TARMAN of lucerne onto the normal dry farming sequence of fallow, wheat, and to a similar introduction by NEJAT BERKMEN of esparsette. Both use the legumes for three years and then continue fallow, wheat. TARMAN (1972) reported that on the same soil with the same seed, fertilizers, and care, the increase in wheat yield was 90%. BERKMEN employing esparsette under the same conditions achieved an increased harvest of 1070 Kg/ha or around 50%.

The selection of crops from the economic viewpoint in north-east Brazil must have reference to a purposeful sequence in the rotation. Thought must also be given to whether perennial crops should be established, or crops with a shorter season that ripen with normal rainy seasons. Perennial crops include cotton, esparsette, lucerne and others; annual include millet, maize, and beans.

It must also be tested whether perennial intercrops increase yields and protect the soil, and whether annual mixed crops such as maize and beans can develop under the special weather conditions, and favour the retention of fertility.

How strongly the present weather of the area can affect unevenly the crop plants can be seen with the variable success of cotton. Dr. ORIOSWALDO MANQUEIRA principal of the research station Serra Talhada in Pernambuco held the view that cotton plants are good only for three years. In a contribution at the Facenda São Miguel in Rio Grande do Norte state the authors wrote 'in more intensive culture the most favourable duration of cropping seems to be 5 years, the plants to be pulled up before the fall of yield in the sixth year'. Dr. PEDRO PIREI, principal of the research station Serido in Rio Grande do Norte state reported on the basis of long experience that cotton plants in the Serido area should be used only for five years.

The research station at Serido initiated notable attempts to solve the problem of rotation in association with perennial cotton. The cotton was planted in strips of four rows with one metre between the rows, and 0.5 metres between plants in the rows. Equally-wide strips alternating with the cotton were planted with a rotation of maize, beans, and millet. The cotton was used for five years, with the best yield in years 2-4. After the harvest in the fifth the cotton was removed and one of the rotation crops planted, whilst the former rotation strip was planted with cotton. The use of the strips thus changed every five years.

The method of strip sewing has theoretically various advantages. The narrow planting of the cotton gives even in the first year a good thick cover protecting the soil. If the strips on sloping ground run along the contour, and on level ground are at a right angle to the wind, soil erosion is effectively contained. At the same time the rotation problem is resolved. A disadvantage is that the strips make preparation and harvesting with machines difficult. Only a long continued careful field experiment can give a clear picture of the advantages and disadvantages of this method of cultivation for the given conditions of soil and climate.

In many parts of north-east Brazil where the cotton rows are farther apart maize, millet and beans are planted between rows. With these crops this practice is possible only for one year after the cotton has been planted, since later the cotton plants are too large. An exception is opuntia. This can remain as a between-row crop, since it is little affected by shade. The between-row crop is important as a soil protection, in particular in the first year after sowing cotton, when the dangers of soil erosion are particularly high.

The maintenance of fertility, the structure and water capacity of arable soils presents serious problems for the farmer of north-east Brazil. On one hand he must maintain the correct humus balance in the soil so that the structure is not lost, and the absorptive capacity for water and nutrients lowered. On the other hand he must secure the mineral supply of the plants without increasing unequally the water needs.

The break-down of humus is promoted in warm climates so that the humus content of the soil diminishes rapidly. If the farmer does not succeed in maintaining the humus level of the soil, the ability of the soil to produce crops is rapidly reduced so that some form of shifting cultivation has to be adopted. Then for years the soil is left unused or used prudently for pasture. Beneath its protective steppe flora the soil slowly regenerates and can be used after several years once more for arable.

If the land is in permanent use then the farmer must maintain the humus balance. In the climate of north-east Brazil this is difficult since the soil is always warm and when sufficient moisture is at hand an active soil fauna rapidly breaks down the humus. Mulching helps to provide organic material, but at the same time helps provide the right conditions for humus break-down. Mulching is therefore not a satisfactory method of maintaining the humus balance. An addition of organic material by dunging is only possible to a limited extent.

To what extent green manure, the ploughing-in of a crop, at best a legume, can promote humus building in north-east Brazil must be established by tests in various areas. It is possible that with fodder crops such as esparsette or lucerne the last cut should be ploughed in and not used for feed. Since the soil under perennial fodder crops is already improved it is to be anticipated that green manure would very favourably influence the humus balance.

Provision of nutrient for crops can be achieved by carefully measured additions of artificial fertilisers. The extent to which trace elements must also be provided is usually recognised from growth phenomena of the crop plants, but is best ascertained by field trials.

A special problem of mineral fertiliser in dry areas is nitrogen yield. This must be very carefully measured. Too high a nitrogen addition stimulates the vegetation growth of the plants, and extends the growth period. With a short rainy season and aridity this can lead to a water shortage for the crop, and to a poor harvest or even failure.

Care of the growing crop has two main tasks. First the soil between the rows must be kept loose so that rain water can be taken up and so that in rainless periods nothing should be lost by evaporation. Second the weeds must be combatted. The wild plants need for growth the same nutrients and water as the cultivated. Because of natural selection in relation to the weather conditions present, they have mostly a higher resistance to aridity and a greater facility to extract moisture. In dry periods the weeds may survive whilst the crops suffer. Wild plants are often a good protection against erosion but at the same time offer cultivated plants a dangerous concurrence in their demands for water and nutrients. In arid areas the saying of Turkish farmers is particularly apposite 'You must kill the son of the field (the weeds) so this step-son (the wop plant) wins life.'

Because of the complexity of rainfall conditions and their range of variation no firm advice for farming in the whole area can be given. The relevance of any advice must be tested locally. With weather changing so rapidly from place to place and from period to period it is not possible in a short period of time to obtain reliable results on the methods required not so much as to bring in record harvests in good years, but to achieve reasonable harvest in bad years.

What is required is research into the time and nature of soil working and seed bed preparation, the containing of wind and water erosion, the maintenance of the humus balance, the time and nature of sewing or planting, a purposeful rotation, the best fertilisers and the care of various crops.

It would be sensible if when various research programmes are developed to deal with the various problems, they are all carried through at all the research stations in north-east Brazil. Furthermore the research must be carried on for a number of years, in order to obtain reliable information with respect to the fluctuating weather conditions and also the effect on the soil of the various agricultural processes.

The decisive question is now: are there parts of north-east Brazil where even with well adapted methods the rainfall is so low that arable agriculture is not possible. In order to answer this question it is relevant to consider the views of R.P. AMBROGGI, and I. LANIR on arable husbandry in the Dorok region of Israel. There the average rainfall is 360 mm. In nine years this total was not reached, in eight years it was exceeded. For every ten years the following picture emerges.

One year of complete drought and no harvest
three years of low rainfall and a bad harvest
two years of average rainfall and average harvest
three years of good rainfall and good harvest
one year of heavy rainfall and very good harvest

The author took the view that under these circumstances — dry farming is unsure and economically unsuccessful.

The precipitation conditions for individual areas of north-east Brazil must be substantiated. Investigations should be undertaken at all stations within the four states that have an average annual rainfall of less than 500 mm. Then the following division can be proposed according to the current rainfall total. With a fall of 0-200 mm a harvest failure is to be expected. A bad harvest is associated with a fall of 200-350 mm, a reasonable harvest 350-500 mm, quite good 500-700 mm, good to very good 700-1500 mm, questionable over 1500 mm. This division is selected because in a dry year with less than 200 mm at the Fazenda São Miguel no usable harvest was expected. On the other hand at the research stations Serva Talhada and Serido a well distributed 300 mm was considered sufficient for a moderate cotton crop. Too wet a year leads also to a doubtful harvest. BOULANGER and his fellow worker in 1966 stated for Sâo Miguel 'With rainfalls exceeding 450 mm by the end of May any prediction is practically impossible'.

As an example of how two places with almost-equal mean rainfalls of 480 mm have quite different harvest prospects in a given year because of differences in the actual rainfall received, the figures for Bom Conselho 9°10'N, 36°41'W, altitude 654 mm, in Pernambuco state and Pirangas 9°37'N, 35°46'W, altitude 47 m in Alagoas state are given below.

Bom Conselho

harvest

decade	harvest failure 0-200 mm	bad 200-350	moderate 350-500	quite good 500-700	good to very good 700-1500	dubious over 1500
1935-44	6	0	1	2	0	1
1945-54	2	0	0	2	6	0
1955-67 (−4)	4	1	0	3	1	1
sum	12	1	1	7	7	1
average	4.3	0.3	0.3	2.4	2.4	0.3

Piranhao

decade	harvest failure 0-200 mm	bad 200-350	moderate 350-500	quite good 500-700	good to very good 700-1500	dubious over 1500
1913-22	0	1	1	7	1	0
1923-32	0	6	2	2	0	0
1933-42	0	0	7	2	1	0
1943-52	0	0	4	6	0	0
1953-62	0	3	4	3	0	0
1963-67	0	1	1	2	1	0
sum	0	11	19	22	3	0
average	0	2.0	3.4	4.0	0.6	0

At Bom Conselho, with 29 years of observations, on average per decade there was to be expected 4.3 harvest failure, 0.3 bad harvests, moderate harvests and dubious harvests, and 2.4 quite good harvests and good harvests. Arable farming in this area is very uncertain and therefore uneconomic.

At Piranhas it is quite different. Rainfall measurements here have gone on for 55 years. No annual rainfall of less than 200 mm was recorded so there were no harvest failures. On average every decade, according to the data, there are to be expected 2 bad, 3.4 moderate, 4.0 quite good, and 0.6 very good harvests. At Piranhao in contrast to Bom Conselho an economic arable farming can be sustained if the farming methods, crop plants and rotation fit the circumstances of the area. This is confirmed by the relative constancy of the rainfall. The maximum rainfall was 784.8 mm in 1914, the minimum in 1928. 202 mm. The difference is 582.8 mm, 120% of the average.

Similar favourable conditions apply at Betania in Pernambuco, 8°17'N, 38°2'W, and at an altitude of 431 m. In 30 years of measurements 494.5 mm were averaged. The variation was only 123.1%. On average in every decade 2.7 bad, 3.3 moderate, 2.7 quite good, and 1.3 good harvests are to be expected, with no harvest failures.

The high variation at many places means that the average rainfall for a decade offers only an approximate guide to anticipated harvests. Serido research station is

Table 16. The different annual rainfalls on average in a decade.

							harvest in mm			
Place			annual mean	obs. years	failed harvest	bad	moderate	quite good	good	dubi
latitude	longtitude	height			0-200 mm	200-350 mm	350-500 mm	500-700 mm	700-1500 mm	>15 mm
Pedra Lavrada 6°45'	36°28'	525 m	283,1	27	4,1	3,3	1,1	1,1	0,4	0
Cabaceiras 7°30'	36°17'	390 m	252,4	54	4,1	3,2	2,2	0,4	0,2	0
Bom Conselho 9°10'	36°41'	654 m	480,3	29	4,1	0,7	0,3	2,4	2,1	0,3
Riacho Fundo 6°33'	36°30'	500 m	302,3	31	3,2	4,2	0,6	1,6	0,3	0
Pau Ferro 8°57'	40°44'	385 m	344,8	21	2,9	3,2	1,4	1,4	1,0	0
Arizona 8°40'	40°58'	500 m	383,6	28	2,9	1,4	2,5	3,2	0	0
Recanto 5°52'	36°17'	400 m	417,5	31	2,9	1,6	2,3	1,3	1,9	0
Desterro 7°17'	37°6'	590 m	321,8	43	2,8	4,0	1,6	0,9	0,7	0
Picui 6°31'	36°22'	450 m	313,0	51	2,7	3,9	2,4	0,8	0,2	0
Sitio Novo 8°48'	38°24'	400 m	347,0	23	2,6	1,8	3,9	1,3	0,4	0
Pixore de Baixo 5°47'	36°36'	122 m	423,2	29	2,4	2,8	2,1	1,0	1,4	0,3
Bodocongo 7°32'	35°59'	350 m	333,8	33	2,4	2,4	3,6	1,2	0,3	0
Barra de Santa Rosa 6°43'	36°4'	440 m	324,5	35	2,3	4,6	1,7	1,1	0,3	0
Caraubas 7°43'	36°41'	460 m	359,6	35	2,3	3,4	3,1	0,6	0,6	0
Currais Novos 6°16'	36°31'	350 m	373,6	51	2,2	2,9	2,2	2,0	0,8	0
Lajes 5°42'	36°15'	198 m	389,2	50	2,2	3,0	2,4	1,6	0,8	0
Sao Rafael 5°48'	36°55'	71 m	451,1	43	2,1	1,9	2,1	1,9	2,1	0
Sao Joao do Cariri 7°24'	36°32'	445 m	386,4	55	2,0	3,3	2,4	1,6	0,7	0
Tara 8°44'	36°52'	586 m	392,7	25	2,0	2,8	2,0	2,0	1,2	0
Jofely 7°4'	36°4'	624 m	340,4	32	1,9	4,4	2,5	1,2	0	0

harvest in mm

ce itude	longtitude	height	annual mean	obs. years	failed harvest 0-200 mm	bad 200-350 mm	moderate 350-500 mm	quite good 500-700 mm	good 700-1500 mm	dubious >1500 mm
uador 57'	36°43'	500 m	392,1	32	1,9	2,8	1,9	2,8	0,6	0
o Jose do Egito 28'	37°17'	575 m	427,1	52	1,9	2,7	1,5	2,7	1,2	0
aja 54'	37°50'	355 m	362,0	27	1,9	2,6	3,3	2,2	0	0
avata 13'	35°34'	447 m	393,0	27	1,9	1,5	4,0	2,6	0	0
dro Avelino 31'	36°23'	97 m	395,4	32	1,9	0,9	5,6	1,0	0,6	0
ledade 4'	36°22'	560 m	375,8	51	1,8	3,7	2,5	1,4	0,6	0
ueirinhas 38'	40°7'	500 m	316,7	18	1,7	3,9	3,3	1,1	0	0
xoto 43'	37°32'	431 m	346,6	31	1,6	4,2	2,6	1,6	0	0
eimadas 12'	35°28'	4 m	406,3	31	1,6	3,9	1,9	1,3	1,3	0
trolandia 4'	38°18'	282 m	407,3	31	1,6	3,6	1,6	2,3	1,0	0
) 52'	38°29'	290 m	401,1	25	1,6	2,0	3,6	2,8	0	0
lhada Real 2'	40°1'	345 m	462,3	26	1,5	3,1	2,3	1,2	1,5	0,4
gadinho 6'	36°51'	410 m	417,8	27	1,5	2,6	1,8	3,7	0,4	0
lem Sâo Francisco 46'	38°58'	305 m	406,8	47	1,5	2,3	3,2	2,3	0,6	0
onso Bezerra 30'	36°30'	80 m	471,7	33	1,5	2,1	2,4	2,7	1,2	0
rgalheira 24'	36°35'	330 m	476,9	40	1,5	1,8	2,5	2,8	1,5	0
o Tome 58'	36°4'	175 m	384,4	31	1,4	2,9	4,1	1,6	0	0
polito 27	37°13'	230 m	489,1	31	1,3	3,2	1,3	1,6	2,6	0
a Maria da Boa Vista 48'	39°50'	452 m	459,2	54	1,1	3,0	2,2	1,9	1,9	0
o Joao do Tigre 4'	36°52'	616 m	413,6	30	1,0	4,7	2,3	1,0	1,0	0

139

Table 16. Continued

			harvest in mm						
Place			annual mean	obs. years	failed bad harvest 0-200 mm	bad 200-350 mm	moderate 350-500 mm	quite good 500-700 mm	good 700-1500 mm	dut >1. mn
latitude	longtitude	height								
Cabrobo 8°30'	39°19'	350 m	469,8	52	1,0	2,7	2,1	2,5	1,7	0
Jua 3°52'	39°53'	180 m	495,5	31	1,0	1,6	2,6	2,9	1,9	0
Petrolina 9°23'	40°30'	376 m	396,2	53	0,8	4,0	3,6	0,9	0,8	0
Floresta 8°46'	38°58'	305 m	456,1	51	0,8	2,7	2,6	2,7	1,2	0
Sume 7°39'	36°56'	510 m	471,4	29	0,7	1,7	3,4	2,1	2,1	0
Olivedo 6°59'	36°15'	545 m	465,5	33	0,6	2,7	3,0	2,1	1,5	0
Curatis 4°59'	40°16'	380 m	436,6	34	0,6	3,2	2,6	2,6	0,9	0
Serra Caiada 6°6'	35°42'	110 m	498,6	43	0,5	2,6	3,3	2,6	0,9	0,2
Santa Cruz 6°14'	36°1'	240 m	489,3	53	0,4	3,0	3,4	2,1	0,9	0,2
Jeritaco 8°23'	37°38'	445 m	486,5	27	0,4	2,2	3,7	2,2	1,5	0
Parambu 6°14'	40°43'	470 m	482,7	33	0,3	1,3	3,6	3,9	0,6	0
Airi 8°42'	38°12'	361 m	464,0	31	0,3	2,9	3,2	2,6	1,0	0
Boa Vista 7°16'	36°14'	490 m	408,8	44	0,2	4,6	2,7	1,8	0,7	0
Sao Caetano 8°19'	36°9'	552 m	480,8	22	0	1,8	4,5	2,7	0,9	0
Pirinhas 9°37'	37°46'	47 m	482,9	55	0	2,0	3,6	3,8	0,6	0
Santo Antonio 5°51'	40°21'	420 m	453,0	33	0	2,4	4,6	2,7	0,3	0
Itagua 6°57'	40°21'	540 m	468,6	33	0	2,7	3,0	3,6	0,6	0
Poco de Pedra AC 6°58'	40°20'	530 m	488,2	33	0	2,4	3,0	3,9	0,6	0
Delmiro Gouveia 9°23'	37°59'	256 m	497,4	29	0	3,1	2,4	3,1	1,4	0
Betania 8°17'	38°2'	431 m	494,5	30	0	2,7	3,3	2,7	1,3	0

140

again an example. In the decade 1930-39 by reference to the rainfall totals there were two failures, 2 bad harvests, 3 moderate, 1 quite good, and 2 good. In the decade 1960-69 in contrast there was neither failure nor poor harvest, but one moderate, three quite good and six good.

In order to help judge which stations with up to 500 mm rain can have economic-farming, table 10 was prepared, giving for individual stations the number of years with 0-100 mm, 101-300 mm, 800-1000 mm and over 100 mm rainfall. For the dry years the table also shows how many followed in sequence. Also the average rainfall is given and the variation. Table 16 shows for the same stations how often in 10 years on average rainfalls of 0-200, 200-350, 350-500-700, 700-1500 and above 1500 mm are recorded.

The table is ordered according to the number of years in a decade with a mean annual rainfall of less than 200 mm of rain, in which according to our reckoning the harvest fails.

At the head of the table is Pedra Lavada where in 27 years of observation and an average rainfall of 283.1 mm it is anticipated that on average in every 10 years there will be 4.1 harvest failures, 3.3 bad harvests, 1.1 moderate, and only 1.1 quite good, plus 0.4 good harvests.

As the table shows for six places, in a decade, on average only 0.5 to 0.2 years occur with less than 200 mm, and seven further places receive generally no such small rainfall. In the assessment if arable husbandry is possible, a year with 200-350 mm must be associated with a bad harvest.

This can be clarified with the region São João do Tigre in Pernambuco state. The station recorded as a mean of 30 years 413.6 mm rainfall. On average in a decade only one year with less than 200 mm was observed, but 4.7 years with 200-350 mm. Here only 10% failed harvests are to be anticipated, but 47% bad harvests and this is not compensated for by 23% moderate and 10% quite good harvests. A test and a comparison of the figures in tables 10 and 11 makes possible an assessment of the possibilities of arable husbandry in districts with less than 500 mm, so long as rainfall observations have been made for a number of years.

Also districts with annual rainfalls not much above 500 mm can be uneconomic for arable husbandry if the variation in rainfall totals in individual years is too great.

In other districts cultivation can be carried on with succes even if a drought year is to be expected as in 1958.

The view expressed in the publication of the Deutschen hygrologischen Mission with Sudene 'Grundlagen für das Bewasserungsproject Araras' that the natural rainfall in the basin of the Rio Araras and the Araras reservoir allows no regular arable husbandry is too one-sided, seen from the viewpoint of irrigation and without reference to the possibilities of a modern rational dry farming.

As a scale for rainfall data for the reservoir the figures for Reriutuba 4°10'N, 40°35'W, at an altitude of 148 m with an average rainfall of 911.0 mm from 54 years observation, and for Araras 4°14'N, 40°28'W, at an altitude of 100 m with 797.6 average rainfall from 29 observation years were used.

At both places the rainfall in 1958 was under 200 mm, and at Riutube also in 1913. With these rainfall amounts and distribution economic farming appears possible, and according to the rainfall data available that is also true for many parts of north-east Brazil. There are however two pre-conditions. First there must be knowledge of a rational dry-farming system tested under the given weather and soil

141

On average for a decade of mean observed rainfall

harvest

failed harvest	bad	moderate	quite good	good to very good	dubious
0-200 mm	200-350 mm	350-500 mm	500-700 mm	700-1500 mm	over 1500 mm
0.4	0.6	0.7	2.0	5.9	0.4

			Araras		
0.3	0.3	1.2	2.7	4.8	0.7

conditions, and this experience won in testing the system must be applied by the farmers. Secondly there must be available a deep soil of good structure and high water-bearing capacity.

An unescapable need in the overcoming of the occasional dry years is storage, and this applies to both food and fodder. In many parts of the world with unreliable weather conditions farmers sell their surpluses after a good harvest only when they are sure that the next harvest will meet their needs. To secure sufficient fodder for cattle it can be stored in a barn or as green fodder in silos.

In the village Vila Santa Cruz between Petrolina and Araripina 60 km from Ouricuri we visited a farmer who on the advice of a veterinary surgeon had filled an earth silo with chopped maize and sugar cane (Fig.18).

Fig. 18. Farmer at Vila Santa Cruz filling soil silo with chopped maize straw and sugar cane.

Fig. 19. Reservoir at Santa Luzia in Paraiba.

Fig. 20. Excavated pond lined with plastic sheet to collect and store rainwater during the rainy season at the research station at Araripe (plateau 810 m).

143

The decisive difficulty for the permanency of the rural population is the shortage of water for man and beast that results from a sequence of dry years. In a few undertakings, attempts have been made to store more in the rainy period in collection ponds or in listerns. This would help in an unusually long dry season, but would fail in a sequence of dry years.

In many regions reservoirs are built with state help (Figs.13, 20), but in some of these the water is saline, as LENZ reported in 1934/35.

More recently SUDENE in collaboration with a German geilogical mission has attempted to solve the problem of water provision in a dry area. The mission works in the region of west Pernambuco and north Bahia. It was reported that test programmes have practically demonstrated the avilability of deep water of good quality to a limited extent. A large part of the study area can be provided with drinking water even in extremely dry years.

It is important to establish the extent to which small irrigation schemes can be built near to villages. If every farmer in the dry area could have half a hectare of irrigable land with fruit and vegetables for his own use in dry years and also fodder, then he could survive economically a sequence of dry years. If the farmers are provided with irrigable land it is essential that they are instructed in irrigation practice, to avoid waste of water and the salinisation of land.

Small scale irrigation has the advantage that a relatively small area will provide security for a large number of farmers, without running the risk of severe over production in the area.

In contrast large scale irrigation, in which only a small number of farmers work, can achieve a production, given the favourable thermal conditions, that will rapidly exceed the needs of the farmers and of the district if no large town is in the vicinity.

An example of an irrigation project is Sume in Paraiba. When complete it will total 400 hectares. In 1972 120 hectares were ready and divided, and a further 160 hectares were to be ready and handed over to farmers in the same year. In 1972, in the main, tomatoes, bananas and beans were planted. Earlier much of the food in the area was imported, but already in 1972, although only a third of the final area was in use, considerable quantities were exported outside the district.

Summary

1. Temperatures in the whole of north-east Brazil are evenly high, and in most districts near the optimum for warmth-loving plants.

2. Throughout north-east Brazil a wet season alternates with a dry season. As the Sudene has shown the rainy season changes from south to north-east by north-west, and then again south.

3. The average annual rainfall is relatively high for a dry area. Of 425 stations only 60 had a rainfall under 500 mm, whereas 71 received over 1000 mm.

4. In the dry year 1958 only a small harvest per hectare was obtained, and also the actual crop area was reduced. In four states the total harvest in cotton was 39.5% that of a good year, beans 15.6%, manioc 52% and maize 17.5%.

5. In 1958 39 stations recorded rainfalls of 0-100 mm, 181 were under 300 mm. At 74% of all rainfall-recording stations seven or more dry months with 0-20 mm were recorded.

144

6. In the whole region the annual total varies markedly. As a measurement of the variation the difference of maximum and minimum falls were considered as a percentage of the mean.

7. In many districts more than one dry year can follow in sequence.

8. In dry years humidity is considerably decreased. The dry air induces higher plant-transpiration despite the poorer provision of water.

9. Climatic and weather diagrams give a good representation of the weather conditions of a region.

10. The arable lands of north-east Brazil are at risk to wind and water erosion. Correct soil preparation, the use of mulches, and planting of wind-breaks are effective counter measures. Research by Dr. MOACIR B. DE FREITAS shows that water erosion can be contained even on slopes of 12%.

11. In combatting drought a good water-storage capacity in the soil is important. The success of modern dry-farming turns on the possible extensive storage of rain water in the soil during the fallow year.

12. If the soil is to retain its facility to absorp nutrients, its water capacity and its resistance to erosion, then despite the climatic difficulties the humus balance must be maintained.

13. For the preparation of the soil in dry areas a goose-foot cultivator with trailer disc-harrow is suitable. Where soil preparation coincides with the beginning of the rainy season sewing should take place at the same time in order that neither time nor moisture are lost. Both tasks can be affected today with one apparatus consisting of a cultivator, drill and harrow.

14. On sloping ground the seed and plant rows should be along the contour.

15. The rotation should maintain fertility and be suitable for present weather conditions.

16. The possibilities of maintaining a humus balance must be developed. The artificial fertilisers must meet the needs of individual crops. Nitrogenous fertilisers must be used with care in order that the vegetative development of the plants is not prolonged, extending the growth period and increasing water needs.

17. In order to be able to give firm advice to the practical farmers the following problems must be solved: a) when and with what apparatus is the soil best prepared for the anticipated rains? b) when and how are the individual crops to be planted? c) what steps must be taken against water and wind erosion? d) how can the humus level in the soil be maintained? e) what form should artificial fertilisers take in composition and quantity?

18. Since at most places the amount of rain, and the length of the rainy season varies from year only long continued research can provide an answer to the various problems. Particularly good results can be expected if a field research programme according to a uniform plan is carried out at all the experimental stations in the dry region of north-east Brazil.

19. Weeds be combatted since they are the strongest competitors for water and nutrients suffered by the crop plants.

20. To test in which areas with an average rainfall of under 500 mm an economic arable husbandry is possible table 16 was constructed. It gives for each of 60 stations the average annual rainfall, the rainfalls that are to be expected on average year by year in a decade ranked at -200 mm, 200-350 mm, 350-500 mm, 500-700 mm, 700-1500 mm, and over 1500 mm, and thus the extent of the variability. In conjunction with table 10 this gives the possibility to test whether

agriculture can be economic at a given station. Even where high mean rainfalls occur agriculture can be uneconomic because of a frequent sequence of dry years.
21. In extreme drought conditions, as in 1958, the farmers must provide for the loss of harvest by storage of food and fodder from previous years.
22. Provision of water for farmers and their stock during a sequence of dry years is a precondition for a permanently settled population. To meet this need Sudene and a German geological mission successfully carried out a test-bore programme in order to find and provide supplies of salt-free water.
23. It is desirable where possible to provide for each farmer an irrigation area of 0.5-1. hectares. On this area the farmer could grow basic necessities for his family and fodder for his stock even in drought years. By this means his economic existence would be generally secured.

Zusammenfassung

'Möglichkeiten und Aussichten der Trockenlandwirtschaft in den Gebieten mit unsicheren Niederschlagsverhältnissen Nordostbrasiliens'.

Das betreffende Gebiet umfaßt 1.546.032 qkm mit 28.160.7 Einwohnern 1971.
1. In ganz Nordostbrasilien ist die Wärme gleichmäßig hoch und ausgeglichen. Sie liegt nahe dem für wärmeliebenden Pflanzen gegebenen Optimum.
2. In dem Gesamtgebiet wechselt jährlich eine Regenzeit mit einer Trockenzeit ab. Im Laufe des Jahres wandert, wie die Sudene gezeigt hat, die Regenzeit von Süden über Nordwesten nach Nordosten und wieder nach Süden.
3. Die mittleren Jahresniederschläge liegen für ein Trockengebiet relativ hoch. Von 425 Stationen haben nur 60 einen mittleren Jahresniederschlag unter 500 mm, dagegen 71 über 1000 mm. (Vergl. Karte 1).
4. 1958 meldeten 39 Stationen 0-100 mm Jahresniederschlag, bei 181 blieb er unter 300 mm. (Vergl. Karte 2). Von 74% aller Regenstationen wurden sieben und mehr trockene Monate mit 0-20 mm Regen gemeldet.
5. Im Trockenjahr 1958 wurde nicht nur eine geringe Ernte je Hektar eingebracht, auch die Anbaufläche war stark vermindert. (Vergl. Tabelle 3a und 3b).
6. Im gesamten Gebiet schwanken die jährlichen Regenmengen sehr. (Vergl. Karte 3).
7. In manchen Gebieten können mehrere Trockenjahre einander folgen. (Vergl. Tabelle 10 und Abb. 2).
8. In Trockenjahren sinkt die Luftfeuchtigkeit erheblich ab. Die trockne Luft zwingt die Pflanzen zu erhöhter Transpiration.
9 Klima oder Witterungsdiagramme geben einen guten Überblick über die Wetterbedingungen eines Gebietes. (Vergl. Abb. 3).
10. In ganz Nordostbrasilien ist der Ackerboden durch Wind- und durch Wassererosion gefährdet. Richtige Bodenbearbeitung, Anwendung von 'Mulch' sowie die Errichtung von Windschutzhecken sind wirksame Schutzmaßnahmen.
11. Im Kampf gegen die Dürre ist eine möglichst gute Wasserspeicherung im Boden wichtig.
12. Trotz der durch das Klima gegebenen Schwierigkeiten, muß alles versucht werden, um die Humusbilanz im Boden aufrecht zu erhalten.
13. Zur Bodenbearbeitung in Trockengebieten erscheint der Gänsefußkultivator

mit angehänger Scheibenegge günstig. (Bild 7 und 8). Erfolgt die Bodenbearbeitung mit Beginn der Regenzeit, sollte die Saat im gleichen Arbeitsgang erledigt werden, damit nicht unnütz Feuchtigkeit verloren geht.

14. In hängigem Gelände sollten Saat- oder Pflanzenreihen quer zum Hang entlang den Konturlinien angelegt werden.

15. Die Fruchtfolge soll die Bodenfruchtbarkeit aufrecht erhalten und muß den jeweiligen Klimabedingungen angepaßt sein.

16. Die Möglichkeiten zur Aufrechterhaltung der Humusbilanz müssen entwickelt werden. Die Mineraldüngung hat den Nährstoffbedarf der einzelnen Kulturen zu decken. Die Stickstoffdüngung ist dabei mit Vorsicht anzuwenden.

17. Um den praktischen Landwirt zuverläßig beraten zu können, sind für alle Trockengebiete Nordostbrasiliens folgende Probleme zu klären: a) Wann und mit welchen Geräten ist der Boden am besten vorzubereiten? b) Wann und wie sind die einzelnen Feldfrüchte zweckmäßig zu bestellen c) Welche Maßnahmen müssen örtlich durchgeführt werden, um den Acker vor Erosion zu schützen? d) Welche Pflanzen können mit Erfolg angebaut werden und wie sind sie in eine Fruchtfolge einzuordnen? d) Wie kann der Humusspiegel des Bodens erhalten werden? f) Wie ist die Mineraldüngung nach Zusammensetzung und Menge der Nährstoffe örtlich am vorteilhaftesten zu gestalten?

18. Bei den an den meisten Orten von Jahr zu Jahr stark wechselnden Regenmengen und der ungleichen Länge von Regen- und Trockenzeit können nur langjährig einheitlich durchgeführte Versuche eine zuverlässige Antwort auf die einzelnen Fragen ergeben. Einen besonderen Erfolg könnte man erreichen, wenn in allen Versuchsstationen der Trockengebiete Nordostbrasiliens die notwendigen Feldversuche nach einheitlichem Plan durchgeführt würden.

19. Das Unkraut muß, da es Wasser- und Nährstoff-Konkurrent der Kulturpflanze ist, sorgfältig bekämpft werden.

20. Um zu prüfen, in welchen Gebieten mit einem mittleren Jahresniederschlag von unter 500 mm ein wirtschaftlicher Ackerbau möglich erscheint, wurde Tabelle 16 aufgestellt. Zusammen mit der Tabelle 10 erlaubt sie zu überschlagen, ob in einem bestimmten Gebiet der Ackerbau rentabel sein kann.

21. In extremen Trockenjahren wie in 1958 muß der Bauer durch eine laufend und konsequent durchgeführte Vorratswirtschaft die Minderernte überwinden.

22. Die Sicherung der Wasserversorgung von Mensch und Tier auch bei eine Folge von Trockenjahren ist die Voraussetzung für eine dauernde Seßhaftigkeit der Landbevölkerung.

23. Wo es irgend möglich ist, wäre es wünschenswert, für jeden Bauern, Wasser zur Bewässrung von 0,5-1,0 ha zu beschaffen. Damit wäre seine wirtschaftliche Existenz auch in Trockenjahren gesichert.

Resumo

'Possibilidades e perspectivas da agricultura de áreas sêcas nas regiões com condições incertas de precipitação no Nordeste do Brasil'.

A região em questão abrange 1.546.032 km^2 com 28.160.878 habitantes em 1971.

1. Em todo o Nordeste do Brasil a temperatura é uniformemente elevada e equilibrada. Situa-se perto do ótimo das plantas termófilas.

2. Na região total uma época de chuvas alterna-se com una época de sêca. Como mostrou a Sudene, a época de chuvas migra do sul sôbre o noroeste ao nordeste e novamente ao sul.

3. As médias de precipitação anual situam-se relativamente alto para umafregião sêca. De 425 estações apenas 60 têm uma precipitação inferior a 500 mm, porém 71 acima de 1000 mm (comp. mapa 1).

4. Em 1958, 39 estações notificaram -100 mm de precipitação anual, em 181 esta permaneceu abaixo de 300 mm (comp. mapa 2). De 74% de tôdas as estações pluviométricas foram notificados sete ou mais meses com -20 mm de chuva.

5. No ano da sêca de 1958 não se verificou apenas uma colheita reduzida por hectar, mas também a área cultivada foi fortemente reduzida (comp. tabelas 3a e 3b).

6. Na região total as chuvas anuais variam muito (comp. mapa 3).

7. Em algumas regiões podem ocorrer diversos anos de sêca subseqüentes.

8. Em anos de sêca, a umidade atmosférica decresce consideràvelmente. O ar sêco obriga as plantas a uma transpiração elevada.

9. Diagramas de clima e de tempo dão uma boa imagem das condições do tempo de uma região (comp. Fig.3).

10. Em todo o Nordeste do Brasil, o solo agrário está sujeito à erosão eólia e hídrica. Conveniente tratamento do solo, aplicação de 'mulch', assim como erguimento de sebes para aparar o vento são medidas protetoras eficazes.

11. Na luta contra a sêca, um bom armazenamento de água no solo, na medida do possível, é importante.

12. Apesar das dificuldades oferecidas pelo clima deve ser feito tudo para manter o balanço humoso no solo.

13. Para o tratamento do solo em regiões sêcas parece ser conveniente o cultivador 'pé de ganso' com rastelo de disco pendente (Figs. 7 e 8). Se o tratamento do solo ocorre no início da época chuvosa, a semeadura deve ser realizada na mesma operação, a fim de evitar perda de umidade.

14. Em terrenos inclinados, as linhas de semeadura ou plantio devem ser dispostas perpendicularmente à inclinação, ao longo das curvas de nível.

15. O afolhamento deve mänter a fertilidade do solo e deve ser adaptado às respectivas condições climáticas.

16. As possibilidades da manutenção do balanço humosos devem ser desenvolvidas. A adubação mineral deve cobrir as necessidades em nutrientes das diversas culturas. A adubação de nitrogênio deve ser aplicada aqui com cuidado.

17. Para poder-se aconselhar seguramente o agricultor prático, devem ser esclarecidos os seguintes problemas para tôdas as regiões sêcas de Nordeste do Brasil: a. Quando e com que ferramentas o solo é preparado da maneira mais adequada? b. Quando e como os diversos produtos agrícolas devem ser cultivados convenientemente? c. Que medidas devem ser tomadas localmente para proteger o terreno contra a erosão? d. Que plantas podem ser cultivadas eficazmente, e como devem ser incluidas no afolhamento? e. Como o teor de humo do solo pode ser mantido? f. Como a adubação mineral é executada localmente de maneira mais eficaz em composição e quantidade?

18. Devido às grandes variações das quantidades de chuva de ano para ano e à duração desigual das épocas de chuva e de sêca, apenas experiências realizadas uniformemente durante anos podem fornecer uma resposta segura às diversas questões. Um succeso especial conseguir-se-ía, se em tôdas as estações experimen-

tais das regiões sêcas do Nordeste do Brasil fossem executados os necessários ensaios de campo segundo um plano uniforme.

19. As ervas daninhas, sendo concorrentes de água e nutrientes para as plantas de cultura, devem ser combatidas cuidadosamente.

20. Para verificar em que regiões com precipitação média anual inferior a 500 mm uma agricultura econômica parece ser possível, foi elaborada a tabela 16. Juntamente com a tabela 10 permite avaliar, se em determinada região a agricultura pode ser lucrativa.

21. Em anos extremamente sêcos, como 1958, o agricultor deve superar a colheita inferior através de u,a economia de armazenamento, executada constante- e conseqüentemente.

22. A garantia do suprimento de água para homem e animal também no caso de uma seqüência de anos de sêca é a condição prévia para um estabelecimento duradouro da população rural.

23. Sempre que possível seria desejoso conseguir água para irrigação de 0,5-1, ha para cada agricultor. Assim a sua existência econômica seria garantida também em anos de sêca.

REFERENCES

BANCO DO NORDESTE DO BRASIL 1968. Manul de Estatisticas do Nordeste Fortaleza-Ceara.
BANCO DO NORDESTE DO BRASIL 1971. A Agricultura do Nordesto. Fortaleza-Ceara.
BANCO DO NORDESTO DO BRASIL 1972. Estatisticas Básicas do Nordeste, Extrativa Vegetal, (Manuscript). Fortaleza-Ceara.
BECK, VON K. 1972. Run-off Bewässerung in der Negev Israëls. Z. für Bewässerungswirtschaft 7.
BOULANGER, I., D. BIRCH, D. PINHEIRO & C.V. FARIA 1966. Flutaçoes da Produçao Algodeiro Mocó. Sudene, Recife.
CHEPIL W.S. 1958. Soil conditions that influence wind erosion. USDA, T.B. 1158, Washington.
COLASAN, UMRAN & EMIN 1960. Türkye Iklimi. (Klima der Türkei) Ankara.
CÖLASAN, UMRAN & EMIN 1962. Sidetli yağis tekerrür analizleri. (Starkregen und Analyse ihres Vorkommens), Ankara.
CHRISTIANSEN-WENIGER F. 1922. Der Energiebedarf der Stickstoffbindung durch Knöllchenbakterien. Centralblatt für Baketriologie Parasitenkunde und Infektionskrankheiten. 58: 1/3.
CHRISTIANSEN-WENIGER, F. 1932. Die anatolische Luzerne und ihr Anbau. Pflanzenbau. 9: 1.
CHRISTIANSEN-WENIGER, F. 1937. The importance of soil erosion for the intensification of field husbandry in Turkey. Herbage Reviews. 5.
CHRISTIANSEN-WENIGER, F. 1938. Orta Anadolu kuru ziraati. (Die Trockenlandwirtschaft im Zentralanatolien), Ankara.
CHRISTIANSEN-WENIGER, F. 1938. Tabii Gübrenin Ehemmiyeti. (Die Bedeutung natürlicher Dünger), Ankara.
CHRISTIANSEN-WENIGER, F. 1939. Anatolian Lucerne. Herbage Reviews. 7: 2.
CHRISTIANSEN-WENIGER,F. 1943. Erfolgreiche Landwirtschaft in Trockengebieten. Ber. d. Landw. Forschungsanstalt Pulawy.
CHRISTIANSEN-WEINIGER, F. 1959. Die Landwirtschaft von Jemen. Ber. über Landwirtschaft, 37: 681-708.
CHRISTIANSEN-WENIGER, F. 1964. Gefährdung Anatoliens durch Trockenjahre und Dürrekastrophen. Ztschrft. f. ausländische Landwirtschaft 3.
CHRISTIANSEN-WENIGER, F. 1966. Israëlische Landwirtschaft Wasserbedarf, Wasserbeschaffung und Wasserausnutzung. Z. f. Bewässerungswirtschaft 2.

CHRISTIANSEN-WENIGER, F. 1970. Ackerbauformen im Mittelmeerraum und nahen Osten, dargestellt am Beispiel der Türkei. Frankfurt.

CHRISTIANSEN-WENIGER, F. 1971. Bodenbearbeitung. Handbuch der Landwirtschaft und Ernährung in den Entwicklungsländern. 2: 126-142.

CHRISTIANSEN-WENIGER, F. & O. TOSUN 1939. Die Trockenlandwirtschaft im Sprichwort des anatolischen Bauern. Ankara.

CHRISTIANSEN-WENIGER, F. & P. LEHMANN 1943. Ein Wetterfilm für die Landwirtschaft. Forschungsdienst 16.

EVENARI, M., L. SHANAN & N. TADMOR 1966. Die Landwirtschaft der Negev-Wüste in Vergangenheit und Gegenwart. *Nova Acta Leopoldina* 176: 149-170.

FINCK, A. 1963. Tropische Böden. Paul Parey

FINCK, A. 1971. Fruchtbarkeit tropischer Böden. Hdb. d. Landwirtschaft und Ernährung in den Entwicklungsländern. 2: 99-125.

FREITAS, M.C. DE. Versuche über Wassererosion und Fruchtfolge sowie Bodenbearbeitung in Pesqueira 1951-1968. Freundlicher Weise zur Verfügung gestellte Versuchergebnisse.

HORN, V. 1965. Weideverhältnisse in den Gebieten Vorderasiens. Schriftenreihe des Tropeninstituts Universität Giessen, G. Fischer, Stuttgart, 99-113.

INSTITUTO BRASILEIRO DE GEOGRAFIA 1970. Concentraçao Agricola Segundo as Micro-Regiôes Homogêneas. Area de 'Atuâçao do IPEANE. Abcar.

JUNG, L. 1964. The influence of the stone cover on run-off and erosion on slate soil. I.A.S.H. comission of land erosion 53: 143-153.

JUNG, L. 1965. Bodenerosion durch Wasser und ihre Bekämpfung. Hdb. f. Landschaftspflege und Naturschutz. 2: 288-303.

JUNG, L. & W. ROHMER 1971. Bodenerosion und Bodenschutz. Hdb. d. Landwirtschaft und Ernährung i.d. Entwicklungsländern. 2: 81-98.

KNAPP, R. 1965. Weide-Wirtschaft in Trockengebieten. Schriftenreihe des Tropeninstituts, Univ. Giessen, G. Fischer, Stuttgart.

KÖPPEN, W. 1931. Grundriss der Klimakunde.

LENZ, Fr. 1935. Zur Limnologie der niederschlagsarmen Gebiete Nordost-Brasiliens. Hydrobiologische Anstalt K.W.I. Plön.

MANGUEIRA, O.B. 1971. Taxa de alogamia na cultura do algodoeiro 'Moco'. Instituto de Pesquisas Agronomicas, Bol. Téc. 50.

MANGUEIRA, O.B., J.T. PEREIRA & A.P. DANTAS 1970. Vantagens da consorciaçao na cultura do algodoeiro 'Moco'. *Ebenda Bol. Téc.* 48.

MANSHARD, W. 1971. Agrarraum der Tropen. Hdb. d. Landwirtschaft und Ernähung i.d. Entwicklungsländern. 2: 26-43.

MINISTERIO DE AGRICULTURA, Escritório de Meteorologia 1970. Normais climatólogicas (Area do Nordeste do Brasil) Periode 1931-1960. Rio de Janeiro.

DEUTSCHE GEOLOGISCHE MISSION IN BRASILIEN 1970, Phänomene der Grundwasserversalzung im Zentrum des Trockengebietes von Nordostbrasilien. Recife.

MISSAO HIDROLOGICA ALEMA 1970. Resultados das pesquisas pedologicas de uma area ao oeste do açude Araras-Ceara. Recife.

MISSAO HIDROLOGICA ALEMA 1971. Grundlagen für das Bewässerungsprojekt Araras. Recife.

MISSAO HIDROLOGICA ALEMA 1971. Verdunstung freier Wasserflächen im Einzugsgebiet des Rio Acaru/Ceara. Recife.

SCHMIDT-LORENZ, R. 1971. Die Böden. Hdb. d. Landwirtschaft und Ernährung i.d. Entwicklungsländern. 2: 44-80.

SILBERSCHMIDT, K., M. VICENTE, M. ENGELHARDT & A. NORONHA 1969. Influence of an excessive substrate moisture level on the germination of rice seeds (*Oryza sativa* L.. *Arq. Inst. Biol. S. Paulo* 36 (2): 89-98.

SILBERSCHMIDT, K., M. VICENTE M. ENGELHARDT & A. NORONHA 1969. Substrate moisture levels for germination testing of some agricultural seeds. *An. Acad. Brasil. Cienc.* 41 (4).

SILBERSCHMIDT, K., M. VICENTE, M. ENGELHARDT & A. NORONHA 1966. Water requirements for optimum germination in corn (zea mays). *Arq. Inst. Biol. S. Paulo* 33 (3): 95-112.

MINISTÉRIO DA AGRICULTURA UND MINSITÉRIO DO INTERIOR 1969. Mapa explotor-io-reconhecimento do solos. Estado do Rio Grande do Norte 1968 und Estado de Pernambuco.

SOUZA REIS A.C. DE & D. DE LIMA 1970. Contribuiçaoao estudo do clima de Pernambuco. Recursos vegetais de Pernambuco. Recife.

SOBRINHO, J. VASCONCELOS 1971. As regiôes naturais do Nordesto o meio e a civilizaçâo. Recife.

SUPERINTENDENCIA DO DESENVOLVIMENTO DO NORDESTE 1963. Normais climato-logicas da área da Sudene.

SUPERINTENDENTIA DO DESENVOLVIMENTO DO NORDESTE 1967. Dados pluviométri-cos mensais 'In natura' 1-3.

TARMAN, O. 1972. Yembitkileri, Cayir ve Mer'a kültürü. Ankara Univ. Ziraat Fak. Yayinl. 464.

WALTER, H. 1955. Klimagramme als Mittel zur Beurteilung der Klimaverhaltnisse für ökologische, vegetationskundliche und landwirtschaftliche Zwecke. *Ber. d. deutsch. Botan. Ges.*. 68: 331-344.

WALTER, H. 1956-57. Klima-Diagramme als Grundlage zur Feststellung von Dürrezeiten. Wasser und Nahrung 1.

Address of Author: Prof. Dr. F. CHRISTIANSEN-WENIGER, Borby-Hof, Post 2331 Barkelsby.